T0340041

Microlepidoptera of Europe

Volume 1

Cees Gielis

Pterophoridae

Editorial foreword

The Lepidoptera are considered to be one of the best known insect orders in Europe and non-specialists may be surprised to learn that we still have an incredible lack of knowledge of these animals even within this part of the World. This is particularly true for the so-called Microlepidoptera. The species composition of the micro-moth fauna is unsatisfactorily known in many areas, particularly towards the south. Data on the bionomics are frequently not available and therefore we cannot well assess environmental threats to these insects. Information on distribution is still grossly incomplete and numerous records are simply based on misidentifications. All these problems are at least to a certain degree caused by the lack of comprehensive identification literature which is a great impediment for naturalists who may contemplate initiating studies on Microlepidoptera. The only publication project dealing with the fauna of an extensive area is the series Microlepidoptera Palaearctica. However, these volumes are more or less revisional monographs and not concise identification guides.

Recognizing these problems, the well known Natural History publisher Peder Skou in the late eighties decided to initiate a new series dealing with European Microlepidoptera. The main scope was the production of identification guides including figures of the adults (colour photographs or paintings) and of genitalia (drawings or photographs) for identification of species, and giving an overview of important data available on bionomics and distribution. Dr. Erik J. van Nieukerken from the National Museum of Natural History in Leiden was asked to assume the job as editor-in-chief from which, however, he stepped down in early 1994 due to many professional duties. The first extensive contacts to potential authors as well as the first stage of editing volume 1 are his merits and we owe him special gratitude. Aware of the urgent need for a series like **Microlepidoptera of Europe** as an impetus for lepidopterology on our continent we finally accepted to act jointly as editors-in-chief, with the arrangement that we would get the necessary support from Leif Lyneborg as a technical editor. The concept of the series was reorganized and the editorial board extended. We are greatly indebted to the publisher, the technical editor and the members of the editorial panel for their continued interest in the series, and their full support.

This first volume of **Microlepidoptera of Europe** deals with the plume moths (Pterophoridae), an easily recognizable family which, however, still harbours a lot of problems with regard to species-level taxonomy. Cees Gielis presents a rather conservative view, considering a number of recently described taxa as synonyms. While recognizing that some of his decisions may proove controversal, we trust that the book will be an invaluable basis for future research in this field.

Cees Gielis manuscript was more or less finished in the early ninethies, and after some editorial difficulties we are now able to present it to the public. In its original concept this series was not planned to include the European part of the former Sovjet Union; this is the reason why that area is not covered in this volume.

The present volume should not be taken as an inchangeable standard for **Microlepidoptera of Europe**. The series is still under development, and we hope to be sufficiently flexible to edit a series of identification books which reflect continously developing concepts on how to deal most appropriate with European Microlepidoptera.

Microlepidoptera are a diverse group of insects, and we want to adjust every single volume to the specific demands that may arise in relation to the group dealt with. We also want to give the individual author(s) the choice to include special contributions on the group in question (e. g. distribution maps, treatments of larval cases, or include neighbouring areas like N. Africa).

The publication of volume 1 of **Microlepidoptera of Europe** is the first important step on a long road which should finally lead to a set of comprehensive books including all families of Microlepidoptera.

P. Huemer, O. Karsholt & L. Lyneborg

Microlepidoptera of Europe

Volume 1

EDITED BY

P. HUEMER, O. KARSHOLT & L. LYNEBORG

Cees Gielis

Pterophoridae

Apollo Books

Stenstrup

1996

Editors: Peter Huemer, Innsbruck; Ole Karsholt, Copenhagen; and Leif Lyneborg, Copenhagen.

Text composed by: Futura DTP, Frederiksberg.

Colour photography by: Geert Brovad, Copenhagen.

Printed by: Litotryk Svendborg A/S.

This publication should be cited as:
Gielis, C., 1996. Pterophoridae. – *In* P. Huemer, O. Karsholt and L. Lyneborg (eds): Microlepidoptera of Europe 1: 1-222.

ISBN 87-88757-36-6
ISSN 1395-9506

Contents

Abstract

The Pterophoridae of Europe (excluding the former Soviet Union) and the Canary Islands and Madeira are treated. All species occurring in this area are included and most of them illustrated in full colour. All species are briefly redescribed mentioning the main characteristic features of the imagines, and the genital structures. A description of the life cycle is given, and also a note on the distribution. A distribution catalogue has been constructed. It is based on own collection data and verified references. The male and female genitalia are illustrated with line drawings. A checklist of hostplants is given.

The following new synonyms have been established: *Capperia sequanensis* Gibeaux, 1990 is synonymized with *Capperia loranus* (Fuchs, 1895); *Hellinsia alpinus* (Gibeaux & Picard, 1992) is synonymized with *Hellinsia carphodactyla* (Hübner, [1813]); and *Merrifieldia inopinata* Bigot, Nel & Picard, 1993 is synonymized with *Merrifieldia malacodactylus* (Zeller, 1847).

Introduction

Historical notes

The first author treating the plume moths in the binominal nomenclature is Linneaus (1758) in his "Systema Naturae", Xth edition. He treated the species in Alucitae with the following diagnosis: "Alis digitalis fissis ad basin". In "Fauna Svecica" (Linnaeus, 1761), other species were briefly described.

Subsequent authors like Scopoli (1763), [Denis and Schiffermüller] (1775), and Haworth (1811) described several new species. In the period from 1796 to 1834 Hübner published his "Sammlung Europäischer Schmetterlinge". This work greatly improved the knowledge of the Lepidoptera, but later gave rise to numerous discussions on the status of many of the treated species, which are briefly described and sometimes poorly illustrated. Discussions analyzing this work are given by Duponchel (1840a, 1840b, 1844) and Zeller (1841). Many of Hübner's species were considered to be synonyms of already described species.

In the late part of the 18th century, it was recognised that at least the Alucitidae (Orneonidae) forms a distinct family. Latreille separated this group as a family in 1796. The division into genera of the remainder of the plume moths was unchanged until 1825, when Hübner published his "Verzeichniss bekannter Schmetterlinge".

In 1841 and 1852, Zeller revised the known world fauna of plume moths, but did not make any essential changes in the already proposed generic system. Wallengren (1862) treated the Scandinavian species and made a finer differentiation of the genera, but left the basic arrangement unchanged.

Tutt (1907) in his very accurate study of both larvae and adults, reached the conclusion that the Pterophoridae should be divided into three subfamilies: Agdistinae, Platyptiliinae and Pterophorinae. He also created a number of new genera and subgenera, mainly for the European fauna. His studies are published in: "A natural history of British Lepidoptera" volume V.

While Tutt was working on the British fauna, Meyrick worked on the microlepidoptera of the entire world, and described many new species of Pterophoridae, publishing his contributions in the pterophorid parts of the "Genera Insectorum" (1908) and "Lepidopterorum Catalogus" (1913b). These publications are examples of the "state of the art" at that time. Meyrick underestimated the importance of the work of Tutt and synonymized many of his taxa. Others have built on Meyrick's work, thus T.B. Fletcher, working on the fauna of the British Empire, and Lord Walsingham, who gave great attention to North and Central America. In North America, Fernald, Barnes, Lindsey, Lange and McDunnough are to be mentioned and in the Pacific region and Japan, Zimmerman, Yano and Gates Clarke. Recently, interest in this family has grown considerably, especially in Europe. Numerous publications on the taxonomy and biology of the species have been by Arenberger, Bigot, Buszko, Gibeaux, Nel and others.

General remarks

Within the Lepidoptera, the family of the Pterophoridae is distinct. The reason for this is clear, as most species show a cleft in the forewing and a double cleft in the hindwing. These characters differentiate them from all other Lepidoptera. Only the family Alucitidae (former Orneodidae) have cleft wings, but with six lobes per wing. This character however, does not apply to the entire family, as the subfamily Agdistinae has uncleft wings. This cleft wing-shape gives the wing a feather-like appearance, which is reflected in popular names of this family: Plume moths (English), Federmotten (German), Fjäder-

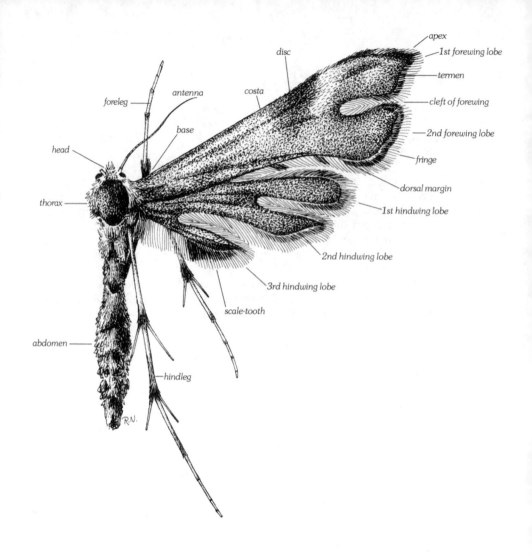

Fig. 1. Sketch of *Buszkoina capnodactylys*, showing main
morphological terms. (R. Nielsen del.).

mott (Swedish), Fjermøl (Danish) and Vedermot (Dutch). These feathers, or lobes as
they are called in this publication, are margined by long fringe-hairs. The fringe-hairs
are differentiated in paler and darker parts. In the fringes, well developed scales lined up
in a row, or forming a scale-group or a scale-tooth may be present.

The wings are narrow. At rest the moth folds and rolls its wings in such a way, that only
a small surface remains exposed, making the specimen hard to detect. The way the
wings is folded seems related to the subfamilies (Wasserthal, 1970). At rest the wings are
extended at right angles to the body, giving the moth a "T"-like appearance.

The basic wing venation (Figs 2-6) is uniform, except for the secondary loss of veins in
certain genera and a shift in origin of some other veins, viz. Cu1 and Cu2. Differences
are especially present in the configuration of the radial veins, and these characters are
used in the generic classification. The presence of one or two veins in the third lobe of
the hindwing was earlier used as a character in the division into subfamilies. However,

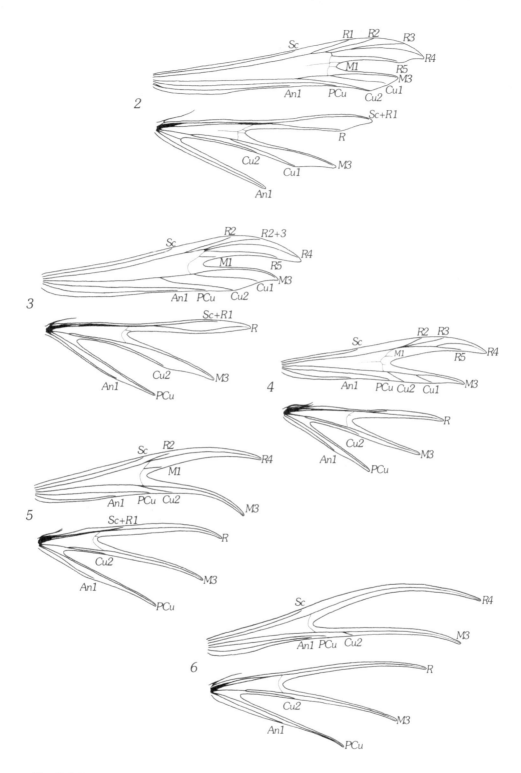

Fig. 2-6. Wing venation in some pterophorid genera. – 2, *Stenoptilia*; 3, *Oidaematophorus*; 4, *Pselnophorus*; 5, *Porrittia*; 6, *Pterophorus*. (After Buszco, redrawn).

phylogenetic studies on the genera have not confirmed the advantage of this character, as it resulted in the formation of paraphyletic groups.

The markings on the forewing are reduced in the Agdistinae to a small number of dots at the costa, at the margin of the naked field, and faint spots at the dorsum. Throughout the family spots at or before the base of the cleft are present. The remaining patterns are more or less related to the subfamily and/or generic grouping of the species. The *Platyptilia* group of genera has a remarkable costal triangle, and in the *Oxyptilus* group transverse brown markings margined by narrow white lines is a characteristic feature. In the *Stenoptilia* group of genera dots and lines before the cleft and in the forewing lobes are present, combined with small scale groups in the terminal fringes of the forewing. In the *Platyptilia* genus group a cluster or row of well developed scales may be present in the fringes of the termen or more commonly the dorsum, in addition to a scale-tooth at the dorsum of the third hindwing lobe. In the Pterophorinae the markings are limited to a more or less well developed number of spots at the costa, the dorsum, near the base of the cleft and in the separate lobes.

A character present in all species belonging to this family, and present in both the males and the females, is the rows of specialized scales on the underside of the hindwing. These scales have been referred to as androconial scales (in German *Duftschuppen*), but their presence in the females makes the status of these scales doubtful.

The head (Figs 7-10) is in general broad, with appressed scales, and a variable number of erect scales in the collar region. The frons (Fig. 7) may show a conical extension of the head. The eyes are spherical and laterally positioned on the head. Ocelli are absent. The maxillary palpi are absent. The labial palpi have three segments with appressed scales, sometimes with small hair-brushes along the segments, and are porrect or upcurved. Between the palpi a rather well developed tongue is present. The antennae are thread-like, with appressed scales and short ciliation. The first segment is enlarged. The antennal length varies from half to the entire forewing length.

The thorax has a compact, cuboid structure. Dorso-lateral shoulder plates or tegulae are present. The thoracic segments are clothed in appressed scales.

The plume-moths have long, slender abdomens and long and slender legs. The fore legs are the shortest, the hind legs by far the longest. The tibiae are longer than the coxae. The hind tibiae possess a double pair of spurs. These spurs may be of equal or unequal length, and the relative length of the proximal and distal spur pair may vary as well. At the base of the spur pairs a scale brush may be present, as for example in the genus *Oidaematophorus*. The five tarsal segments show considerable differences in size, the basal segment being the longest. At the apex of the fifth tarsal segment is a pair of small claws.

On the margins of the segments of the abdomen small scale brushes may be present; they are often paired and dorsally placed. This feature is seen in some species of the genus *Stenoptilia* and *Pterophorus*. In the genus complex *Oxyptilus* a lateral hair-brush is present along the ninth segment. Both on the dorsal and ventral side of the abdomen line-shaped markings or patterns may occur.

The terminology of the male genitalia appears in Figs 17, 39, 46, 66, 76, and of the female genitalia in Fig. 186.

The division into subfamilies of the Pterophoridae in Europe is based on the following characters:

– Agdistinae. The members of this subfamily are without clefts in the wings. The colour of the wings is dull grey to ochreous grey. On the distal third of the forewing there is a distinct triangular "naked field". (The naked field is a triangular field containing fewer cover scales. It plays a role in wing folding). The apex of this triangle is near the transverse vein.

– Pterophorinae. The members of this subfamily have their forewings cleft once.

In the tropical areas of the world two more subfamilies are recognized:

– Ochyroticinae. This subfamily has uncleft wings, as in Agdistinae, but has no "naked field" on the forewing. Instead a longitudinal, brigth orange-yellow pattern interrupted by silvery-white is seen.
– Deuterocopinae. In the members of this subfamily the forewing is cleft two or three times.

The sequence of the genera appearing in this book is based on the phylogenetic studies by Gielis (1993). The sequence of the species in the genera has no phylogenetic basis.

The Pterophoridae is a very widespread family with a world-wide distribution. In mountainous areas considerable altitudes may be reached. The species occur under very different ecological conditions. One example of the diversity is found in the group of species in the genus *Agdistis* which live in salt-marshes along the coasts and in desert-like habitats in southern Spain. Another example is the species of *Stenoptilia* living at high altitudes in the Alps and also occurring in Iceland and Greenland under arctic conditions. Other species live in shady woods among herbaceous vegetation, or in clearings in woodland. Also extremely wet habitats like moorlands, wet meadows and sides of rivers, brooks and ditches are utilised by certain species, and drier habitats like heaths and road-sides add to the range.

The biology of the different species reflects the diversity of their habitats. A major group of larvae feeds in the rootstocks and stems of the hostplants, and hibernate therein. Other species feed on the foliage. A number of species hibernate in stems and feed in spring and/or summer on the leaves, flowers and seeds. A peculiar species in this respect is *Adaina microdactyla* which lives in the stem of *Eupatorium* in the winter generation and on the leaves and flowers in summer. *Buckleria paludum* is remarkable as it feeds on the leaves of *Drosera*. The major hostplant groups are the Asteraceae, Labiatae, Gentianaceae and Rosaceae.

The larvae of Pterophoridae are easily recognised. They tend to possess a dense pattern of setal hairs. The setae are long and often flattened distally. In a number of cases these setal hairs produce a sticky fluid which seems to act as a protective trap against parasites. In general the setal hairs are arranged in groups, which are usually forming four longitudinal rows on the larva. The long-haired larvae feed externally on the host plants. A typical feeding pattern is creating fenestration of the leaves, as the larva eats the underside of a leaf and leaves the upper cuticle untouched. Another biological group of larvae have short setal hairs. These larvae tend to be root or stem borers. In case of *Adaina microdactyla* the borer causes galls in the stem of the host plant. The long-haired larvae are usually marked by a mixture of green and brown spots and lines, which do not form a characteristic common pattern. The boring larvae are yellow-white without a pattern; some species are reddish brown. The first-instar larva (or ´egg larva´) has significantly shorter and less densely set hairs than the larvae of the second and later instars.

In the text part, a short description of the biology of the species has been given. It is important to realize that these biological data is mostly based on material and publications originating from western and central Europe, as biological data referring to species in the Mediterranean area are very limited. The recent efforts by various French authors in the south of France and on the island of Corsica will hopefully fill in these gaps. The known biology and host plants are separated from those data that are not fully verifiable or dubious.

The pupae tend to have a similar, but less developed, hair pattern as the larvae. In particular the pupae of the borers have poorly developed hairs. A peculiar dorsal structure on the pupae is seen in some representatives of the genera *Platyptilia*, *Stenoptilodes*, *Lantanophaga*, *Paraplatyptilia* and *Stenoptilia*. The significance of this structure is at present uncertain. Externally the imaginal structures appear on the surface of the pupa; especially the head, tongue, palpi and wings are easily observed.

The eggs lack distinguishing characteristics. They are rounded to oval, elliptical, with a slightly flattened top; the surface is smooth to delicately reticulate, especially near the top of the egg; the colour varies from white-yellow and yellow to pale and deep green.

How to collect plume-moths

It requires some rules and standards to establish a collection of plume-moths, or of any other group of insects. The main rule is not to collect more specimens than necessary for establishing the identity of the species. If possible, the specimens collected should be killed and processed right away. This prevents over-collecting, causes no threat to the population's existence, and also ensures beautiful and well set material. Nextly, the specimen must be labelled, so that information with locality, date, altitude, ecological data, method of collecting and collector is attached. This labeling makes the specimen scientifically valuable, and can be used for collecting data for research.

Specimens of Pterophoridae may be collected in a variety of ways, depending on the behaviour of the species. On mountain slopes in the Alps specimens may fly after being disturbed. These specimens tend to re-settle on the vegetation after a short flight, and are easily found again or netted in flight. The vegetation may also be swept, using a relatively stiff net. This method is specially useful in collecting representatives of the genus *Capperia*, which are poor flyers and hardly ever come to light. The netting of species flying at dusk is an obvious method. To facilitate collecting a headlamp or miner's lamp may be used. The moths may be recognized in flight, or found resting on flowers. The use of light to collect Lepidoptera is widely used all over the world. There is a significant difference in results when using lamps with different spectral range. The widely used mercury-vapour lamp has a high content of ultraviolet radiation, and so has the low pressure mercury lamp, but its content of U.V. is less. Actinic lights have a different spectrum as well and are in use with or without a visible light component. Finally the common household bulb may be used with its large proportion of infrared light. Each of these types of lamps has advantages, which are related to the circumstances in which one has to use the lamp. It is noteworthy that some species are attracted to the M.V.-lamp, while other species are only seen at the household bulb. Besides, a number of species is attracted by the M.V.-lamp, but fly to the household bulb to relax on the screen or in the vegetation. Therefore, the author always uses two sheets with different types of lamps next to one another.

I am well aware that private insect collections is a source of much controversy. As a taxonomist, I consider the existence of insect collections to be essential. Without them there would be no detailed knowledge of insects. The pressure on nature, however, and especially in a rural continent like Europe, makes it essential that the insects collected are treated in a scientifically responsible way, and that collecting is kept to a minimum.

The build-up of a collection is greatly influenced by its purpose. Some collectors or museums wish to have a local or regional collection. Others specialize in a family or some other category. This is are visualized in the way a collection is built up and exhibited. In all cases, a systematic approach is essential for an identification of the specimens and an arrangement of the identified specimens. To serve as a guide for arranging a collection, a checklist of the European species is given in the following pages. As stated above no specimen should be left unlabelled.

The preparation of genital slides

It is not always possible to identify insects with certainty on the external characters. Late in the 19th century it was recognized that the genitalia of species closely resembling each other in external characters, may show distinct differences in the genital structure of both the male and female (Hofmann, 1896; Fernald, 1898). Ever since, the examination of the genital structures has been an important part of the description of the species. Major work in this field has been done in the European, especially British, fauna by Pierce & Metcalfe, who produced a series of books treating the Lepidoptera of the British Isles. In North America, similar efforts were done by Barnes & Lindsey. Nowadays the description of the genitalia is an essential part of the description of a species, and illustrating the genitalia is at least as important as illustrating the specimen.

The technique in preparing a genital slide may vary. The procedure described below is the technique used in the main museums of the world and accepted as a general standard. The specimen to be examined is inspected to find out if the abdomen is attached in a natural way to the thorax, or it has been glued to it. In the latter case extreme care has to be used in removing the abdomen from the thorax. The abdomen has to be gently moved up and down and from left to right in order to loosen the abdomino-thoracical junction, without damaging it. After loosening, the abdomen is placed in a tube containing a 10% solution of potassium-hydroxide (KOH). The tube is heated "au-bain-marie" to 100° Celsius. Depending on the size of the abdomen, the hydroxylation of the fat content takes from one to several minutes. The abdomen is then placed in a watch glass containing 30% ethanol. Under a magnifying glass or stereo-microscope the abdomen is gently cleaned for scales by rubbing it with a small brush or the tip of a feather. By far the best to use is the 'pin feather' (1st primary) of a woodcock (*Scolopax rusticola*). When the surface scales are removed, the structures inside the abdomen become visible. The abdomen is then stained in a solution of Chlorazol Black E. This dye gives the chitinous parts a blue-grey colour (in case of overstaining the colour may be reduced by house-hold bleach). The genital structures are then easily recognizable and differentiated from the other abdominal organs. Now the internal structures are subsequently cleaned by gently and carefully dragging out the gut. In case of a male, the genital structures are separated from the abdomen, the remaining fragments removed, and the aedeagus is taken out of the apparatus. In a female genital apparatus, the abdominal segments are separated at the 7th or 8th sternite, depending on the structures in the genitalia. A further rinsing may take place. After the cleaning phase, the genital structures are arranged in position on an object-slide. The abdomen and the genital parts should be placed in a ventro-dorsal position separately, and cleaned in such a way that no detritus or scales remain. The genitalia and abdomen on the object-slide are dehydrated in steps with ethanol 50%, 70% and 100%. Euparal-essence may be added, and embedding takes place in Euparal. Finally both the insect and the genital slide are labelled and numbered correspondingly. On the slide-label is indicated the slide number, the name of the species, the collecting data, the embedding agent and date. The slides take several months to harden. This means they have to be stored horizontally, or have to be hardened in an artificial way. The hardening may be quickened by placing the slides in an oven at 50°-65° Celsius for 24 hours.

Also in use, but less favoured by the author, are embedding agents based on xylene (xylol). The processing of the slides is more critical, and the agents tend to turn yellow after a period of some years, making the interpretation very difficult. Another problem arises with the cristalline hardening of the preparations. The use of Tenebrinta veneziana (natural resin of larch) is less problematic: the slides made in the years 1910-1922 in the USNM are still in a perfect state. In contrast to this is the use of Faure's embedding agent, which eventually dries and in a number of cases has resulted in severe mould on the preparation, making the slides useless for examination.

Key to the European genera of Pterophoridae

1. Wings cleft (Fig. 1) 2
   – Wings not cleft .... *Agdistis* (p. 27)
2. Third lobe of hindwing with 1 vein (Fig. 2) 3
   – Third lobe of hindwing with 2 veins (Figs 3-6) 22
3. Forewing with markings in form of transverse lines on the lobes;
      wings slender, without a termen on the first lobe of the forewing 5
   – Forewing with markings in form of a costal triangle, spots before
      the base of the cleft or in the disc; a well developed termen is
      present in the first lobe of the forewing (Fig. 1) 13
5. Palpi with a hair-brush along the third segment (Figs 9, 10) 6
   – Palpi without a hair-brush along the third segment (Fig. 8) 7
6. Third lobe of hindwing with scale-tooth at apex .... *Oxyptilus* (p. 76)
   – Third lobe of hindwing with scale-tooth at middle .... *Crombrugghia* (p. 80)
7. Valvae with a vesicular lobe at middle (Fig. 96) .... *Buckleria* (p. 75)
   – Valvae without such a lobe 8
8. Valvae flap-like, rounded (Fig. 106) .... *Megalorhipida* (p. 83)
   – Valvae elongate 9
9. Valvae entirely simple, without widenings or spined structures
      (Figs 81, 105) 10
   – Valvae with widened sections and/or spined structures 11
10. Uncus and tegumen small, simple (Fig. 81) .... *Geina* (p. 67)
    – Uncus and tegumen with hooked socii (Fig. 105) .... *Stangeia* (p. 83)
11. Valvae in apical half gradually widened, no basal widening with
       spiny structures (Fig. 84) 12
    – Valvae basally widened, with spiny structures or teeth-like hooks
       directed towards the valve-base (Fig. 85) .... *Capperia* (p. 70)
12. In the male the apex of the ninth sternum is short, blunt, and
       covered with a brush (Fig. 84) ....*Paracapperia* (p. 69)
    – In the male the apex of the ninth sternum is bifurcate
       (Fig. 82) .... *Procapperia* (p. 68)
13. Sacculus simple (Fig. 77) 14
    – Sacculus more or less bilobed 18
14. Valvae with a spine or spines directed towards the dorsum
       (Fig. 77) 15
    – Valvae without spines 16
15. Third lobe of hindwing with an apical scale-tooth .... *Cnaemidophorus* (p. 64)
    – Third lobe of hindwing without a scale-tooth .... *Marasmarcha* (p. 65)
16. Frons at most slightly protruding 17
    – Frons conically protruding for more than twice the eye-
       diameter (Fig. 7) .... *Gillmeria* (p. 42)
17. Third lobe of hindwing with a scale-tooth at apex .... *Buszkoiana* (p. 64)
    – Third hindwing lobe with a scale-tooth at middle .... *Platyptilia* (p. 38)
18. Saccus with bristles on an apical plate (Figs 52, 53)............... *Amblyptilia* (p. 47)
    – Saccus without such bristles 19
19. Saccus forked apically (Fig. 49) .... *Lantanophaga* (p. 44)
    – Saccus simple 20
20. Uncus short and small, reaching beyond margin of the tegumen
       by at most about half of its length. Third lobe of hindwing
       without a scale-tooth .... *Stenoptilia* (p. 49)

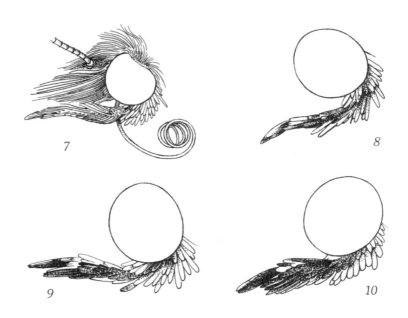

Fig. 7. Head in profile of *Gilmeria tetradactyla*. Figs. 8-10. Palpi of 8, *Geina didactyla*; 9, *Oxyptilus pilosella*; 10, *Crombrugghia distans*. (R. Nielsen del.).

Acknowledgements

During the preparation of this work, I have received generous advise, support, and assistance from many colleagues. I have kindly been permitted to examine numerous collections and have had material at my disposal. I am particularly indebted to E. Arenberger, Vienna, Austria; B.-Å. Bengtsson, Löttorp, Sweden; L. Bigot, Marseilles, France; F. Coenen, Brussels, Belgium; M. Corley, Faringdon, UK; A. Cox, Mook, Netherlands; W. Dierl, München, Germany; J.P. Duffels, Amsterdam, Netherlands; C. Gibeaux, Avon, France; P. Huemer, Innsbruck, Austria; M. Hull, Helsby, UK; R. Johansson, Växjö, Sweden; R. de Jong, Leiden, Netherlands; O. Karsholt, Copenhagen, Denmark; the late F. Kasy, Vienna, Austria; J.C. Koster, Callantsoog, Netherlands; J. Kuchlein, Wageningen, Netherlands; G. Langohr, Simpelveld, Netherlands; S. Löser, Düsseldorf, Germany; J. Lucas, Rotterdam, Netherlands; G. Luquet, Paris, France; J. Nel, La Ciotat, France; E.J. van Nieukerken, Leiden, Netherlands; W.O. de Prins, Antwerp, Belgium; L. de Ridder, Antwerp, Belgium; R.U. Roessler, Karlsruhe, Germany; K. Schnack, Söborg, Denmark; R.T.A. Schouten, Oegstgeest, Netherlands; M. Shaffer, London, UK; S. Yu. Sinev, St. Petersburg, Russia; H.W. van der Wolf, Nuenen, Netherlands; J.B. Wolschrijn, Twello, Netherlands.

Special thanks go to E. Arenberger, who provided me with numerous specimens and genital preparations needed to identify and illustrate the species. His criticism on the first drafts of the manuscript has been a positive stimulus for further work and been of great help. B. Goater, who corrected the final drafts of the English text, is also warmly thanked.

Finally, I am most grateful to the Uyttenboogaart-Eliasen Foundation (Amsterdam, Netherlands) which supported the author with grants for museum visits to London, Paris, Karlsruhe, München and Vienna and for the illustrations.

Checklist of European Pterophoridae

The generic sequence is adapted from the phylogenetic study by Gielis (1993). The species sequence in the genera is morphologically without a phylogenetic basis.

AGDISTINAE Tutt, 1907

Agdistis Hübner, [1825]
Adactylus Curtis, 1834
Agdistes Stephens, 1835 (incorrect spelling)
Adactyla Zeller, 1841 (emendation)
Ernestia Tutt, 1907
Herbertia Tutt, 1907 (synonym & homonym)

1. tamaricis (Zeller, 1847)
    bagdadiensis Amsel, 1949
2. intermedia Caradja, 1920
    hungarica Amsel, 1955
3. bennetii (Curtis, 1833)
4. meridionalis (Zeller, 1847)
    staticis Millière, 1875
    tyrrhenica Amsel, 1951
    prolai Hartig, 1953
5. salsolae Walsingham, 1908
    pinkeri Bigot, 1972
6. delicatulella Chrétien, 1917
    melitensis Amsel, 1954
7. neglecta Arenberger, 1976
8. protai Arenberger, 1973
9. adactyla (Hübner, [1819])
    huebneri (Zeller, 1841)
    delphinensella Bruand, 1859
10. heydeni (Zeller, 1852)
    canariensis Rebel, 1896
    excurata Meyrick, 1921
11. satanas Millière, 1875
    nanus Turati, 1924
    pseudosatanas Amsel, 1951
12. frankeniae (Zeller, 1847)
    lerinsis Millière, 1875
    bahrlutia Amsel, 1955
    fiorii Bigot, 1960
    tondeuri Bigot, 1963
    rupestris Bigot, 1974
13. gittia Arenberger, 1988
14. espunae Arenberger, 1978
15. glaseri Arenberger, 1978
16. bigoti Arenberger, 1976
17. symmetrica Amsel, 1955
18. manicata Staudinger, 1859
    gigas Turati, 1924
    lutescens Turati, 1927
    tunesiella Amsel, 1955

19. paralia (Zeller, 1847)
20. bifurcatus Agenjo, 1952
21. pseudocanariensis Arenberger, 1973
22. hartigi Arenberger, 1973
23. betica Arenberger, 1978

PTEROPHORINAE Zeller, 1841
PLATYPTILIINAE Tutt, 1907

Platyptilia Hübner, [1825]
Platyptilus Zeller, 1841 (emendation)
Fredericina Tutt, 1905

24. tesseradactyla (Linnaeus, 1761)
      fischeri (Zeller, 1841)
25. farfarellus Zeller, 1867
26. nemoralis Zeller, 1841
      saracenica Wocke, 1871
      grafii Zeller, 1873
27. gonodactyla ([Denis & Schiffermüller], 1775)
      megadactyla ([Denis & Schiffermüller], 1775)
      diptera (Sulzer, 1776)
      trigonodactyla (Haworth, 1811)
      farfara Gregson, 1885
28. calodactyla ([Denis & Schiffermüller], 1775)
      petradactyla (Hübner, [1819])
      zetterstedtii (Zeller, 1841)
      taeniadactyla South, 1882
      leucorrhyncha Meyrick, 1902
      doronicella Fuchs, 1902
29. iberica Rebel, 1935
      nevadensis Rebel, 1935
30. isodactylus (Zeller, 1852)
      brunneodactyla D. Lucas, 1955

Gillmeria Tutt, 1905

31. miantodactylus (Zeller, 1841)
32. pallidactyla (Haworth, 1811)
      migadactylus (Curtis, 1827)
      ochrodactyla (Treitschke, 1833)
      marginidactylus (Fitch, 1854)
      nebulaedactylus (Fitch, 1854)
      bertrami (Roessler, 1864)
      bischoffi (Zeller, 1867)
      cervinidactylus (Packard, 1873)
      adustus (Walsingham, 1880)
33. tetradactyla (Linnaeus, 1758)
      ochrodactyla ([Denis & Schiffermüller], 1775)
      dichrodactylus (Mühlig, 1863)
      borgmanni (Roessler, 1880)
      bosniaca (Rebel, 1904)

Lantanophaga Zimmerman, 1958

34. pusillidactylus (Walker, 1864)
    technidion (Zeller, 1877)
    hemimetra (Meyrick, 1886)
    lantana (Busck, 1914)
    lantanadactyla (Amsel, 1951)

Stenoptilodes Zimmerman, 1958

35. taprobanes (Felder & Rogenhofer, 1875)
    brachymorpha (Meyrick, 1888)
    seeboldi (Hofmann, 1898)
    terlizzii (Turati, 1926)
    zavatterii (Hartig, 1953)
    legrandi (Bigot, 1962)

Paraplatyptilia Bigot & Picard, 1986
Mariana Tutt, 1905 (homonym)

36. metzneri (Zeller, 1841)
    bollii (Frey, 1856)

Amblyptilia Hübner, [1825]
Amplyptilia Hübner, [1825] (incorrect spelling)
Amblyptilus Wallengren, 1862 (emendation)

37. acanthadactyla (Hübner, [1813])
    calaminthae Frey, 1886
    tetralicella Hering, 1891
38. punctidactyla (Haworth, 1811)
    cosmodactyla (Hübner, [1819])
    ulodactyla (Zetterstedt, 1840)
    stachydalis (Frey, 1870)

Stenoptilia Hübner, [1825]
Mimaeseoptilus Wallengren, 1862
Mimeseoptilus Zeller, 1867 (emendation)
Mimaesoptilus Snellen, 1884 (incorrect spelling)
Doxosteres Meyrick, 1886
Mimaesioptilus Barrett, 1904 (incorrect spelling)
Adkinia Tutt, 1905
Adkina Yano, 1963 (incorrect spelling)

39. graphodactyla (Treitschke, 1833)
40. pneumonanthes (Büttner, 1880)
    nelorum Gibeaux, 1989
    arenbergeri Gibeaux, 1990
41. gratiolae Gibeaux & Nel, 1990
    paludicola auct., nec Wallengren, 1859

42. pterodactyla (Linnaeus, 1761)
  fuscus (Retzius, 1783)
  fuscodactyla (Haworth, 1811)
  ptilodactyla (Hübner, [1813])
  paludicola (Wallengren, 1859)
43. mannii (Zeller, 1852)
  megalochra Meyrick, 1927
44. veronicae Karvonen, 1932
45. bipunctidactyla (Scopoli, 1763)
  mictodactyla ([Denis & Schiffermüller], 1775)
  hodgkinsonii (Gregson, 1868)
  hirundodactyla (Gregson, 1871)
  ? plagiodactylus Stainton, 1851
  ? serotinus Zeller, 1852
  ? scabiodactyla Gregson, 1871
  ? succisae Gibeaux & Nel, 1991
46. aridus (Zeller, 1847)
  grisescens Schawerda, 1933
  csanadyi Gozmány, 1959
  gallobritannidactyla Gibeaux, 1985
  ? mimula Gibeaux, 1985
  ? picardi Gibeaux, 1986
47. elkefi Arenberger, 1984
48. lucasi Arenberger, 1990
49. annadactyla Sutter, 1988
  annickana Gibeaux, 1989
50. pelidnodactyla (Stein, 1837)
  ? alpinalis Burmann, 1954
  ? bigoti Gibeaux, 1986
  ? gibeauxi Nel, 1989
  ? cerdanica Nel & Gibeaux, 1990
  ? cebennica Nel & Gibeaux, 1990
  ? mercantourica Nel & Gibeaux, 1990
50a brigantiensis Nel & Gibeaux, 1992
  buvati Nel & Gibeaux, 1992
51. reisseri Rebel, 1935
52. hahni Arenberger, 1989
53. millieridactyla (Bruand, 1861)
  saxifragae Fletcher & Pierce, 1940
54. islandicus (Staudinger, 1857)
  borealis (Wocke, 1864)
55. parnasia Arenberger, 1986
56. coprodactylus (Stainton, 1851)
  zalocrossa Meyrick, 1907
  ? pseudocoprodactyla Gibeaux, 1992
57. lutescens (Herrich-Schäffer, 1855)
  arvernicus (Peyerimhof, 1875)
  grandis Chapman, 1908
58. nepetellae Bigot & Picard, 1983
  cyrnea Nel, 1991
59. stigmatodactylus (Zeller, 1852)
  oreodactylus (Zeller, 1852)
60. stigmatoides Sutter & Skyva, 1992

61. zophodactylus (Duponchel, 1840)
    loewii (Zeller, 1847)
    canalis (Walker, 1864)
    semicostata (Zeller, 1873)

Buszkoiana Koçak, 1981
Richardia Buszko, 1978 (homonym)

62. capnodactylus (Zeller, 1841)

Cnaemidophorus Wallengren, 1862
Cnemidophorus Zeller, 1867 (emendation & homonym)
Eucnemidophorus Wallengren, 1881
Euenemidophorus Pierce & Metcalfe, 1938 (incorrect spelling)

63. rhododactyla ([Denis & Schiffermüller], 1775)
    koreana (Matsumura, 1931)

Marasmarcha Meyrick, 1886

64. lunaedactyla (Haworth, 1811)
    phaeodactyla (Hübner, [1813])
    agrorum (Herrich-Schäffer, 1855)
    altaica Krulikowskij, 1906
    tuttodactyla Chapman, 1906
65. fauna (Millière, 1871)
66. oxydactylus (Staudinger, 1859)
    wullschlegeli Müller-Rutz, 1914

Geina Tutt, 1907

67. didactyla (Linnaeus, 1758)
    brunneodactylus (Millière, 1854)

Procapperia Adamczewski, 1951

68. maculatus (Constant, 1865)
69. croatica Adamczewski, 1951

Paracapperia Bigot & Picard, 1986

70. anatolicus (Caradja, 1920)
    tamsi (Adamczewski, 1951)

Capperia Tutt, 1905

71. britanniodactylus (Gregson, 1867)
    teucrii (Jordan, 1869)
72. celeusi (Frey, 1886)
    intercisus (Meyrick, 1930)
73. trichodactyla ([Denis & Schiffermüller], 1775)
    leonuri (Stange, 1882)
    affinis (Müller-Rutz, 1934)

74. fusca (Hofmann, 1898)
    marrubii Adamczewski, 1951
75. bonneaui Bigot, 1987
76. hellenica Adamczewski, 1951
77. loranus (Fuchs, 1895)
    sequanensis Gibeaux, 1990
78. marginellus (Zeller, 1847)
79. zelleri Adamczewski, 1951
80. polonica Adamczewski, 1951
81. maratonica Adamczewski, 1951

Buckleria Tutt, 1905

82. paludum (Zeller, 1841)
    paludicola (Fletcher, 1907)
    dolichos (Matsumura, 1931)

Oxyptilus Zeller, 1841

83. pilosellae (Zeller, 1841)
    bohemanni (Wallengren, 1862)
84. chrysodactyla ([Denis & Schiffermüller], 1775)
    hieracii (Zeller, 1841)
85. ericetorum (Stainton, 1851)
86. parvidactyla (Haworth, 1811)
    obscurus (Zeller, 1841)
    hoffmannseggi (Möschler, 1866)
    maroccanensis (Amsel, 1956)

Crombrugghia Tutt, 1907
Combrugghia Neave, 1939 (incorrect spelling)

87. distans (Zeller, 1847)
    clarisignis (Meyrick, 1924)
    buvati (Bigot & Picard, 1988)
    propedistans (Bigot & Picard, 1988)
    adamczewskii (Bigot & Picard, 1988)
    pravieli (Bigot, Nel & Picard, 1989)
    gibeauxi (Bigot, Nel & Picard, 1990)
    jaeckhi (Bigot & Picard, 1991)
88. tristis (Zeller, 1841)
89. kollari (Stainton, 1851)
90. laetus (Zeller, 1847)
    loetidactylus (Bruand, 1859)
    lantoscanus (Millière, 1883)

Stangeia Tutt, 1905

91. siceliota (Zeller, 1847)
    ononidis (Zeller, 1852)

Megalorhipida Amsel, 1935
Megalorrhipida Amsel, 1935 (incorrect spelling)

92. leucodactylus (Fabricius, 1794)
    defectalis (Walker, 1864)
    congrualis (Walker, 1864)
    oxydactylus (Walker, 1864)
    ochrodactylus (Fish, 1881)
    hawaiiensis (Butler, 1882)
    centetes (Meyrick, 1886)
    compsochares (Meyrick, 1886)
    ralumensis (Pagenstecher, 1900)
    derelictus (Meyrick, 1926)
    palaestinensis Amsel, 1935

Puerphorus Arenberger, 1989

93. olbiadactylus (Millière, 1859)
    hemiargus (Meyrick, 1907)
    dactilographa (Turati, 1927)

Gypsochares Meyrick, 1890

94. baptodactylus (Zeller, 1850)
95. bigoti Gibeaux & Nel, 1990
96. nielswolffi Gielis & Arenberger, 1992

Pselnophorus Wallengren, 1881
Crasimetis Meyrick, 1890

97. heterodactyla (Müller, 1764)
    brachydactyla (Kollar, 1832)
    aetodactylus (Duponchel, 1840)

Hellinsia Tutt, 1905
Leioptilus Wallengren, 1862 (homonym)
Lioptilus Zeller, 1867 (emendation & homonym)

98. inulae (Zeller, 1852)
    coniodactylus (Staudinger, 1859)
99. carphodactyla (Hübner, [1813])
    buphthalmi (Hofmann, 1898)
    inulaevorus (Gibeaux, 1989)
    alpinus (Gibeaux & Picard, 1992)
100. chrysocomae (Ragonot, 1875)
    bowesi (Whalley, 1960)
101. osteodactylus (Zeller, 1841)
    cinerariae (Millière, 1869)
102. pectodactylus (Staudinger, 1859)
    angustus (Walsingham, 1880)
    stramineus (Walsingham, 1880)
    melanoschisma (Walsingham, 1908)
103. distinctus (Herrich-Schäffer, 1855)
    sibericus (Caradja, 1920)
    zermattensis (Müller-Rutz, 1934)

104. didactylites (Ström, 1783)
    scarodactyla (Hübner, [1813])
    icarodactyla (Treitschke, 1833)
105. tephradactyla (Hübner, [1813])
106. lienigianus (Zeller, 1852)
    melinodactylus (Herrich-Schäffer, 1855)
    scarodactylus (Becker, 1861)
    serindibanus (Moore, 1887)
    sericeodactylus (Pagenstecher, 1900)
    victorianus (Strand, 1913)
    hirosakianus (Matsumura, 1931)

Oidaematophorus Wallengren, 1862
Oedaematophorus Zeller, 1867 (emendation)
Ovendenia Tutt, 1905

107. lithodactyla (Treitschke, 1833)
    septodactyla (Treitschke, 1833)
    similidactylus (Dale, 1834)
    phaeodactylus (Stephens, 1834)
    lithoxylodactylus (Duponchel, 1840)
108. rogenhoferi (Mann, 1871)
109. constanti (Ragonot, 1875)
110. giganteus (Mann, 1855)
111. vafradactylus Svensson, 1966

Emmelina Tutt, 1905

112. monodactyla (Linnaeus, 1758)
    bidactyla (Hochenwarth, 1785)
    cineridactylus (Fitch, 1854)
    naevosidactylus (Fitch, 1854)
    pergracilidactylus (Packard, 1873)
    barberi (Dyar, 1903)
    pictipennis (Grinnell, 1908)
113. argoteles (Meyrick, 1922)
    jezonicus (Matsumura, 1931)
    komabensis (Matsumura, 1931)
    menoko (Matsumura, 1931)
    yanagawanus (Matsumura, 1931)
    pseudojezonica Derra, 1987

Adaina Tutt, 1905

114. microdactyla (Hübner, [1813])
    montivola Meyrick, 1937

Calyciphora Kasy, 1960

115. punctinervis (Constant, 1885)
    tyrrhenica (Amsel, 1954)
116. homoiodactyla (Kasy, 1960)
117. adamas (Constant, 1895)

118. acarnella (Walsingham, 1898)
119. albodactylus (Fabricius, 1794)
        xerodactylus (Zeller, 1841)
        xanthodactylus auct. (nec Treitschke, 1833)
        sicula (Fuchs, 1901)
120. xanthodactyla (Treitschke, 1833)
        klimeschi Kasy, 1960
121. nephelodactyla (Eversmann, 1844)
        apollina (Millière, 1883)

Porrittia Tutt, 1905

122. galactodactyla ([Denis & Schiffermüller], 1775)

Merrifieldia Tutt, 1905

123. leucodactyla ([Denis & Schiffermüller], 1775)
        leucodactyla (Hübner, [1805])
        theiodactyla (Hübner, [1825])
        wernickei (Wocke, 1898)
        fitzi (Rebel, 1912)
        dryogramma (Meyrick, 1930)
        tridactylus auct., (nec Linnaeus, 1758)
        tetradactyla auct., nec (Linnaeus, 1758)
124. tridactyla (Linnaeus, 1758)
        fuscolimbatus (Duponchel, 1844)
        icterodactylus (Mann, 1855)
        noctis (Caradja, 1920)
        menthae (Chrétien, 1925)
        phillipsi (Huggins, 1955)
        exilidactyla (Buszko, 1975)
        neli Bigot & Picard, 1989
125. baliodactylus (Zeller, 1841)
        meridionalis (Staudinger, 1880)
126. malacodactylus (Zeller, 1847)
        meristodactylus (Zeller, 1852)
        indocta (Meyrick, 1913)
        subtilis (Caradja, 1920)
        parca (Meyrick, 1921)
        subcretosa (Meyrick, 1922)
        phaeoschista (Meyrick, 1923)
        spicidactyla (Chrétien, 1923)
        rayatella (Amsel, 1959)
        insularis (Bigot, 1961)
        livadiensis (Zagulajev & Filippova, 1976)
        transdanubinus (Fazekas, 1986)
        garrigae Bigot & Picard, 1989
        moulignieri Nel, 1991
        inopinata Bigot, Nel & Picard, 1993
127. semiodactylus (Mann, 1855)
128. hedemanni (Rebel, 1896)
        hesperidella (Walsingham, 1908)

129. chordodactylus (Staudinger, 1859)
   probolias (Meyrick, 1891)
   particiliata (Walsingham, 1908)
130. bystropogonis (Walsingham, 1908)

Wheeleria Tutt, 1905

131. phlomidis (Staudinger, 1870)
132. raphiodactyla (Rebel, 1900)
133. spilodactylus (Curtis, 1827)
   confusus (Herrich-Schäffer, 1855)
134. obsoletus (Zeller, 1841)
   desertorum (Zeller, 1867)
   gonoscia (Meyrick, 1922)
   marrubii (Wasserthal, 1970)
   phlomidactylus (Wasserthal, 1970)
135. lyrae (Arenberger, 1983)
136. ivae (Kasy, 1960)

Pterophorus Schäffer, 1766
Pterophorus Geoffroy, 1762, suppressed (ICZN Op. 228)
Plumiger Valmont-Bomare, 1791 (unavailable)
Pterophora Hübner, [1806], suppressed (ICZN Op. 97)
Pterophora Hübner, 1822
Aciptilia Hübner, [1825]
Aciptilus Zeller, 1841 (emendation)
Acoptilia Agassiz, 1847 (emendation)
Acoptilus Agassiz, 1847 (emendation)
Alucita auct., (nec Linnaeus, 1758) (ICZN Op. 703)

137. pentadactyla (Linnaeus, 1758)
   tridactyla (Scopoli, 1763)
138. ischnodactyla (Treitschke, 1833)
   actinodactyla (Chrétien, 1891)
   eburnella (Amsel, 1968)

Systematic Treatment of the Genera and Species of Pterophoridae in Europe

SUBFAMILY: AGDISTINAE Tutt, 1907

This subfamily contains one genus, *Agdistis* Hübner. The subfamily is characterized by the non-cleft wings, dull grey to ochreous grey colour, and the presence of a "naked field" on the terminal half of the forewing.

Agdistis Hübner, [1825]

Agdistis Hübner, [1825]: 429.
  Type species: *Alucita adactyla* Hübner, [1823]; monotypy.
Adactylus Curtis, 1833: 471.
  Type species: *Alucita adactyla* Hübner, [1823]; original designation.
Agdistes Stephens, 1835; incorrect spelling.
Adactyla Zeller, 1841; emendation.
Ernestia Tutt, 1907: 128.
  Type species: *Agdistis lerinsis* Millière, 1875; monotypy.
Herbertia Tutt, 1907.
  Type species: *Adactyla tamaricis* Zeller, 1847; monotypy. Synonymy: Meyrick (1908).

DIAGNOSIS.– Forewing and hindwing entire, not cleft. Colour grey to grey-brown, with a silver gloss in some species. At the forewing costa, small spots are generally present in the terminal half; also spots along the margin of the forewing fold (see below). Forewing veins R1 present, R2 and R3 separate, M3 and Cu1 stalked. The forewings are rolled in the rest position of the moth. The margins of these folds enclose a sparsely scaled field with a wedge-like shape. The top of this field lies near the discal area and the base consists of the central part of the termen.

MALE GENITALIA.– Valvae, saccular and cucullar processes often asymmetrical. Membraneous folds or processes are occasionally present, and so are spines. The 9th sternite is well-developed, and extended.

FEMALE GENITALIA.– These show a great variation; especially the shape and size of the antrum is characteristic.

DISTRIBUTION.– The genus is known from the Palaearctic region, mainly dominating in the Mediterranean zone, Africa and South Asia. A single species is recorded from North America. The genus seems to be replaced by the genus *Ochyrotica* in South-East Asia, Australia and the Neotropical region.

BIOLOGY.– The larvae on the leaves of Plumbaginaceae, Frankeniaceae, Tamaricaceae and Asteraceae. Pupation on a leaf or at the stem.

REMARKS.– The species within this genus are so alike that identification on external characters is in many cases hardly possible and often unreliable. The best differences are found in the genitalia. These are however often complex and the illustrations are far more informative than the description. However, the main characters are mentioned in the text part.

1. *Agdistis tamaricis* (Zeller, 1847)

Adactyla tamaricis Zeller, 1847: 899.   Pl. 1: 1. Figs 16, 153.
Agdistis bagdadiensis Amsel, 1949: 310.

DIAGNOSIS.– Wingspan 18-27 mm. Colour grey-brown. Along the costa four spots, the terminal two smaller than the basal two. The space between spots 1 and 2 is longer than between 3 and 4. Dorsally from the second spot an additional spot. Along the margin of the fold 3 spots, the terminal being the largest.

MALE GENITALIA.– Valve long, without adherent processes. Aedeagus curved and its apex oblique.

FEMALE GENITALIA.– Ostium with an acute apex. Antrum parabolic narrowing. Two large and stout dorsally directed spines next to the antrum.

DISTRIBUTION.– The Canary Islands and the Mediterranean area, as far north as southern Germany and Strasbourg in France, the Balkan states; extending into Asia Minor, Iran, Afghanistan, Pakistan and China, to the south into Israel and Arabia, to the west again in North Africa.

BIOLOGY.– The moth flies from March to October, in successive generations and depending on the latitude. The hostplants are *Tamarix gallica* L. (Hofmann, 1896; Amsel, 1935b; Hannemann, 1977b) and *Myricaria germanica* Desvaux (Hofmann, 1896; Mitterberger, 1912; Arenberger, 1977; Hannemann, 1977b). In South Germany the larvae hatch in autumn and feed on the plants. After hibernation they pupate in spring. A peculiar habit is the throwing away of the excreta by a swinging movement of the larval body.

2. *Agdistis intermedia* Caradja, 1920

Agdistis benneti var. *intermedia* Caradja, 1920: 88.   Pl. 1: 2. Figs 17, 154.
Agdistis hungarica Amsel, 1955b: 53.

DIAGNOSIS.– Wingspan 24-30 mm. Colour grey-brown. The bold field between the folds paler than the ground-colour. No spots along the costa, along the dorsal margin of the fold 4 spots. The 4th spot closer to the costa than the 3rd spot, occasionally they are confluent.

MALE GENITALIA.– Valvae and costal processes symmetrical. Uncus bilobed, symmetrical. The 9th sternite asymmetrically bidentate.

FEMALE GENITALIA.– Ostium slightly rounded. Antrum almost semi-circular, latero-caudally extended into two spine-like processes (as in *A. bennetii*, but less developed). Apophyses anteriores short.

DISTRIBUTION.– Hungary and Romania, extending into Russia.

BIOLOGY.– The moth flies in June and August. The probable hostplant is *Limonium vulgare serotinum* Reichenb. (=*Statice gmelini* Koch) (Arenberger, 1976a).

REMARKS.– The species is closely related to *A. bennetii*, differing in the genitalia.

3. *Agdistis bennetii* (Curtis, 1833)

Adactyla bennetii Curtis, 1833: 471.   Pl. 1: 3; pl. 15: 1, 2. Figs 18, 155.

DIAGNOSIS.– Wingspan 24-30 mm. Colour and markings as in *A. intermedia.*

MALE GENITALIA.– Valvae asymmetrical. Costal processes wide and separate from valvae. also sacculus well separated. Uncus bilobate, each lobe with 3-4 teeth.

FEMALE GENITALIA.– Ostium simple. Antrum in shape of a straight tube. Proximal margin of the 8th tergite excavated, without apophyses anteriores.

DISTRIBUTION.– Along the coasts of England, Denmark, Germany, Netherlands, Belgium, France, Spain, Italy, Yugoslavia and Albania.

BIOLOGY.– The moth flies from May to September, in two generations. The hostplant (in the Netherlands and England) is *Limonium vulgare* Miller (Janmoulle, 1939; Hannemann, 1977b; Emmet, 1979; Gielis, bred). This plant grows on saltmarshes. Related plants and hybrids of the plant are used in gardens. The larva feeds on the leaves. Pupation occurs on the underside of a leaf.

REMARKS.– The species is generally collected on saltmarsh localities where the hostplant occurs. One specimen has been collected in a garden near cultivated plants, over 100 km from its natural hostplant vegetation. In Denmark the species has a tendency to migrate, and occasional specimens are taken inland in southern England.

4. *Agdistis meridionalis* (Zeller, 1847)

Adactyla meridionalis Zeller, 1847: 898.   Pl. 1: 4. Figs 19, 156.
Agdistis staticis Millière, 1875: 375.
Agdistis tyrrhenica Amsel, 1951b: 103.
Agdistis prolai Hartig, 1953: 67.

DIAGNOSIS.– Wingspan 22-25 mm. Colour grey, speckled with brown scales. A subterminal costal spot; four spots present along the dorsal margin of the fold, the 4th obliquely above the 3rd.

MALE GENITALIA.– Valvae asymmetrical. Left valve widened before the apex, with a costal arm two thirds as long as the valve. Right valve with a saccular process and a costal arm. The 9th sternite asymmetrically cleft.

FEMALE GENITALIA.– Ostium slightly excavated. Antrum twice as long as wide, gradually narrowing. 8th tergite with a small central excavation.

DISTRIBUTION.– Canary Islands, Great Britain, and the Mediterranean area, including North Africa. To the east as far as Cyprus.

BIOLOGY.– The moth flies from April till September. The hostplants are *Limonium vulgare* Miller, *L. binervosum* (G. E. Sm.) Salmon. (Beirne, 1954; Arenberger, 1977; Emmet, 1979), *L. obtusifolium* (Rouy) Erben (Nel, 1991), *L. virgatum* (Willd.) Fourr. (Nel, 1991) and *L. cordatum* Miller (Walsingham, 1908; Bigot & Picard, 1983a). On dry sea cliffs in southern England, where the foodplant grows.

5. *Agdistis salsolae* Walsingham, 1908

Agdistis salsolae Walsingham, 1908: 922.   Pl. 1: 5. Figs 20, 157.
Agdistis pinkeri Bigot, 1972: 224.

DIAGNOSIS.– Wingspan 16-18 mm. Colour grey-brown, speckled with dark scales along the dorsum. At the costa four spots, and three spots along the dorsal margin of the fold.

MALE GENITALIA.– Valvae asymmetrical. Left valve with four membraneous processes, and a simple costal arm. Right valve with four distally directed membraneous processes, and one process directed toward the base of the valve. Aedeagus with numerous small cornuti.

FEMALE GENITALIA.– Ostium almost flat. Antrum very wide, one and a half times as long as wide, slightly narrowing. Ductus bursae with a sclerotised plate. Apophyses anteriores short.

DISTRIBUTION.– Canary Islands.

BIOLOGY.– The moth flies in March and April, in June and July and again in October; probably in three generations. The hostplant is *Salsola oppositifolia* Desfontaines (Arenberger, 1977).

6. *Agdistis delicatulella* Chrétien, 1917

Agdistis staticis var. *delicatulella* Chrétien, 1917: 462.   Pl. 1: 6. Figs 21, 158.
Agdistis melitensis Amsel, 1954a: 52.

DIAGNOSIS.– Wingspan 13-17 mm. Colour pale grey, speckled with sparse dark scales. At the costa three spots, between these spots the fringes are almost white; an additional spot obliquely under the central spot.

MALE GENITALIA.– Valvae almost symmetrical. Left valve with a irregular surface ventrally. The right valve with a small process near the apex. Costal arms long and slender, as long as the valvae. Aedeagus with a long cornutus carrying a thorn-like projection at both ends.

FEMALE GENITALIA.– Ostium oblique. The sclerotised part of the antrum with margins oblique to the longitudinal axis of the ductus bursae, and asymmetrically extended. The apophyses anteriores as long as the 8th tergite.

DISTRIBUTION.– Malta, Corsica.

BIOLOGY.– The moth flies in July, September and October. The hostplant is unknown.

7. *Agdistis neglecta* Arenberger, 1976

Agdistis neglecta Arenberger, 1976a: 62.   Pl. 1: 7. Figs 22, 159.

DIAGNOSIS.– Wingspan 16-21 mm. Colour brown, along the costa and the dorsum paler, to grey-white. At the costa four distinct spots. The distance between spots 3 and 4 shorter than between the others. Dorsally of the second spot an additional spot and also three spots on the dorsal margin of the fold.

MALE GENITALIA.– Valvae with regular excavations ventrally, the left valve ending in a small membraneous process, the right valve rounded apically. Costal arms long and slender, as long as the valvae. Aedeagus with a distinct cornutus, with 3-4 thorn-like processes.

FEMALE GENITALIA.– Ostium slightly excavated. Antrum squarish, weakly sclerotised. Ductus bursae with a sclerotised plate. Apophyses anteriores short.

DISTRIBUTION.– Spain, southern France, Balearic Islands.

BIOLOGY.– The moth flies in May and June. The hostplants are *Atriplex portulacoides* L. (Nel, 1989c; Nel, 1991), *Euphorbia pithyusa* L. (Nel, 1989c) and a *Frankenia* sp. (Arenberger, 1976a).

8. *Agdistis protai* Arenberger, 1973

Agdistis protai Arenberger, 1973c: 650.   Pl. 1: 8. Figs 23, 160.

DIAGNOSIS.– Wingspan 15-17 mm. Colour grey, speckled with dark scales. Without discernable costal spots, only a small dark dash at the apex. Along the dorsal margin of the fold three dark spots.

MALE GENITALIA.– Valvae asymmetrical. Left valve with costal and dorsal processes, the apex bifid. The right valve with costal and dorsal processes which are differently shaped and positioned compared to left valve, and a single stout tip. Costal arms asymmetrical, approximately half as long as the valve.

FEMALE GENITALIA.– Ostium asymmetrically displaced. Antrum proximally obliquely margined. Apophyses anteriores short.

DISTRIBUTION.– Italy, Sardinia.

BIOLOGY.– The moth flies in July. The hostplant is unknown.

9. *Agdistis adactyla* (Hübner, [1819])

Alucita adactyla Hübner, [1819]: t. 7, figs 32-34.   Pl. 1: 9. Figs 24, 161.
Adactyla huebneri Zeller, 1841: 771.
Agdistis adactyla var. *delphinensella* Bruand [in: Laboulbène], 1859: 893.

DIAGNOSIS.– Wingspan 22-26 mm. Colour dark grey. At the costa four dark spots, the basal spots the most disticnt. Along the dorsal margin of the fold three spots.

MALE GENITALIA.– Valvae asymmetrical. Left valve with a club-like apex. Right valve with a more acute apex. The left costal arms club-like, the right gradually narrowing, with thickened apex. Aedeagus with a peculiar hooked appearance.

FEMALE GENITALIA.– Ostium and antrum of a sector-like shape, tapering into the narrower and more straight ductus bursae. The 8th tergite distally with an extra-sclerotised margin and short apophyses anteriores.

DISTRIBUTION.– South-West, Central and Northern-Central Europe. Extending to the east as far as Mongolia and Afghanistan.

BIOLOGY.– The moth flies from mid-June to mid-August. The hostplants are *Artemisia campestris* L. (Hofmann, 1896; Gozmány, 1962; Hannemann, 1977b; Buszko, 1979c; Buszko, 1986) and *Chenopodium fruticosum* L. (Arenberger, 1977). The larvae on the lower parts of the plants. Other recorded hostplants are *Erica cinerea* L. (Sutter, 1991) and *Santolina chamaecyparissus* L. (Sutter, 1991).

10. *Agdistis heydeni* (Zeller, 1852)

Adactyla heydeni Zeller, 1852: 322. Pl. 2: 1. Figs 25, 162.
Agdistis heydeni ssp. *canariensis* Rebel, 1896: 114.
Agdistis excurata Meyrick, 1921: 423.

DIAGNOSIS.– Wingspan 17-23 mm. Colour grey, the dorsal field paler, white-grey. At the costa four spots, and an additional spot dorsally between the 2nd and 3rd spots. Along the dorsal margin of the fold three spots, the terminal spot the largest.

MALE GENITALIA.– Valvae asymmetrical. Left valve basally wide, at one third from base suddenly narrowing and progressing into a slender club-like extension. Right valve long and wide, with a central narrowing. The costal arms nearly symmetrical, slightly more than half as long as the valve.

FEMALE GENITALIA.– Ostium deeply excavated in the 8th tergite. Antrum almost squarish, but slightly tapering. In the ductus bursae a proximal sclerotised segment at two thirds of the length of the ductus. Small central process in both lobes of 7th tergite.

DISTRIBUTION.– Canary Islands, and the entire Mediterranean area.

BIOLOGY.– The moth flies in June and July. In the Canary Islands also from March to May and again in September and October. The hostplants are *Atriplex halimus* L. (Walsingham, 1908; Arenberger, 1977; Bigot & Picard, 1983) and *Stachys glutinosa* L. (Nel, 1991).

REMARKS.– The population in the Canary Islands seems to represent a distinct subspecies, as the specimens differ in the genital structure from European specimens (Ridder, 1986).

11. *Agdistis satanas* Millière, 1875

Agdistis satanas Millière, 1875b: 167. Pl. 2: 2. Figs 26, 163.
Agdistis nanus Turati, 1924: 150.
Agdistis pseudosatanas Amsel, 1951b: 102.

DIAGNOSIS.– Wingspan 15-19 mm. Colour dark grey to black-grey. The fringes in the terminal half of the costa white, with three dark spots, subapically again black. Along the dorsal margin of the fold three spots.

MALE GENITALIA.– Valvae slightly asymmetrical in small details. Left valve a little narrower than the right valve. Costal arms short. Uncus deeply cleft.

FEMALE GENITALIA.– Ostium sinuate, bulged out. Antrum almost squarish, extending into a narrow ductus bursae. The 8th tergite longer than wide. The 7th tergite with an acute, bifid apex.

DISTRIBUTION.– Mediterranean area, to the north as far as southern Germany.

BIOLOGY.– The moth flies from April to September. The hostplants are *Scabiosa candicans* Jordan (Millière, 1875), *S. pyrenaica* All. (Hannemann, 1977b), *Scleranthus* sp. (Hofmann, 1896) and *Limoniastrum monopetalum* Boissier (LHomme, 1939; Arenberger, 1977; Bigot & Picard, 1989).

12. *Agdistis frankeniae* (Zeller, 1847)

Adactyla frankeniae Zeller, 1847: 900.   Pl. 2: 3. Figs 27, 164.
Agdistis lerinsis Millière, 1875b: 168.
Agdistis bahrlutia Amsel, 1955b: 51.
Agdistis fiorii Bigot, 1960b: 201.
Agdistis tondeuri Bigot, 1963a: 9.
Agdistis paralia rupestris Bigot, 1974: 85.

DIAGNOSIS.– Wingspan 15-27 mm. Colour grey, speckled with brown scales. At the costa four distinct spots. An additional spot dorsally to the 2nd spot. Along the dorsal margin of the fold three spots, the terminal spot being the largest.

MALE GENITALIA.– Valvae bilobed, with a costal arm longer than the valvae; a small doubled process dorsally at the base of the second lobe. Tegumen bilobate, with small socii. Uncus long, with a circular apex, bearing two small lateral processes. The 9th sternite with a simple stout process.

FEMALE GENITALIA.– Ostium slightly excavated. Antrum at left laterally deviating and oblique; the ductal part gradually narrowing. Apophyses anteriores slender and as long as the papillae anales.

DISTRIBUTION.– Canary Islands, the entire Mediterranean area, and extending into Romania and Russia.

BIOLOGY.– The moth flies from April till September. The hostplants belong to the genus *Frankenia* (Frankeniaceae) (Chrétien, 1891; Walsingham, 1908; Arenberger, 1977; Nel, 1989c). Bigot & Picard (1983a) mention *Limonium minutum* (L.) Fourr.

13. *Agdistis gittia* Arenberger, 1988

Agdistis gittia Arenberger, 1988b: 65.   Pl. 2: 4. Figs 28, 165.

DIAGNOSIS.– Wingspan 22-23 mm. Colour grey-brown, speckled with numerous white scales. At the costa four dark spots. The central field of the forewing without any markings.

MALE GENITALIA.– Valvae bilobed, with long, blunt costal arms, which are longer than the valvae. Valvae as in *A. frankeniae*. Tegumen bilobed, the socii smaller than in *A. frankeniae*. The uncus long and slender, apically cleft. The 9th sternite with a stout apical process.

33

FEMALE GENITALIA.– Ostium and antrum as in *A. frankeniae*. In the ductus bursae a scler-otised U-shaped plate, which is unilaterally covered with numerous spines.

DISTRIBUTION.– Southern Spain.

BIOLOGY.– The moth flies in May. The hostplant is unknown.

14. *Agdistis espunae* Arenberger, 1978

Agdistis espunae Arenberger, 1978: 75.   Pl. 2: 5. Figs 29, 166.

DIAGNOSIS.– Wingspan 22-25 mm. Colour grey, speckled with dark scales. At the costa four spots, the greatest distance occurs between the 2nd and 3rd spots. Along the dorsal margin of the fold three spots.

MALE GENITALIA.– Valvae as in *A. frankeniae*, except for the longer and more acute costal arms. Tegumen bilobed. Socii longer than half the stalk length to the apex of the uncus. Uncus elongate, with two small lateral processes. The 9th sternite with a single stout, rather long process. A cornutus present in the vesica of the aedeagus.

FEMALE GENITALIA.– Ostium deeply excavated, and the antrum of different shape and translocated to the left (much further as than in *A. frankeniae*). Apophyses anteriores as long as the papillae anales.

DISTRIBUTION.– Southern Spain.

BIOLOGY.– The moth flies in June and again in September and October. The hostplant is unknown.

15. *Agdistis glaseri* Arenberger, 1978

Agdistis glaseri Arenberger, 1978a: 77.   Pl. 2: 6. Figs 30, 167.

DIAGNOSIS.– Wingspan 24-26 mm. Colour grey, sparsely speckled with brown scales. The terminal half of the costal fringes white, and four spots at the costa. Along the dorsal margin of the fold three spots.

MALE GENITALIA.– Valvae bilobed, acutely terminated; costal arm longer than valvae. Tegumen bilobed. Socii stout and half as long as the uncus. The apex of the uncus wide and flattened. The apex of the 9th sternite bifid.

FEMALE GENITALIA.– Ostium excavated. Antrum rather wide proximally, directed toward the left, and becoming gradually narrower. Apophyses anteriores slender and as long as the papillae anales.

DISTRIBUTION.– Southern Spain.

BIOLOGY.– The moth flies in September and October. The hostplant is unknown.

16. *Agdistis bigoti* Arenberger, 1976

Agdistis bigoti Arenberger, 1976b: 7.   Pl. 2: 7. Figs 31, 168.

DIAGNOSIS.– Wingspan 24-26 mm. Colour grey. Along the costa an ill-defined brown line with two spots. Three ill-defined spots along the dorsal margin of the fold.

MALE GENITALIA.– Valvae as in *A. frankeniae*, except for the processes at the junction between the lobes. The costal arms acute, but shorter. Tegumen bilobed. The socii stout and long. The apex of the uncus rounded, with small processes at the proximo-lateral margin. The 9th sternite very deeply cleft, forming pair of long extensions.

FEMALE GENITALIA.– Ostium wide, and asymmetrically deeply excavated. Antrum one and a half times as long as wide, sharply narrowing. Apophyses anteriores as long as the papillae anales.

DISTRIBUTION.– Greece: Crete.

BIOLOGY.– The moth flies from May to August. The hostplant is unknown.

17. *Agdistis symmetrica* Amsel, 1955

Agdistis symmetrica Amsel, 1955b: 23.   Pl. 2: 8. Figs 32, 169.

DIAGNOSIS.– Wingspan 19-22 mm. Colour grey. At the costa four spots, and along the dorsal margin of the fold three spots. The spots are ill-defined. This species has almost pectinate antennae.

MALE GENITALIA.– The valvae as in *A. frankeniae*. Tegumen bilobed. The socii half as long as the stalk length to the apex of the uncus. The uncus cleft at the apex, and with two small proximo-lateral processes. The apex of the 9th sternite weakly bifurcate.

FEMALE GENITALIA.– Antrum wide, asymmetrically, but gradually, narrowing into the short ductus bursae. Ostium almost smooth, slightly excavated.

DISTRIBUTION.– Malta, Tunisia.

BIOLOGY.– The moth flies in August. The hostplant is unknown.

REMARKS.– The species is closely related to *A. glaseri*, but differs in the shape of the tip of the uncus.

18. *Agdistis manicata* Staudinger, 1859

Agdistis manicata Staudinger, 1859: 258.   Pl. 3: 1. Figs 33, 170.
Agdistis gigas Turati, 1924: 149.
Agdistis lutescens Turati, 1927: 337.
Agdistis tunesiella Amsel, 1955b: 55.

DIAGNOSIS.– Wingspan 21-27 mm. Colour pale grey to grey. At the costa four spots, the terminal spot hardly visible. Three spots along the dorsal margin of the fold; the 2nd and 3rd are often conspicuous.

MALE GENITALIA.– Valvae bilobed, the basal lobe bifid at apex, whereas the apex of the second lobe is rounded. The costal arms long, slender and pointed. The tegumen is bilobed. The socii about as long as the stalk of the uncus. The apex of the uncus large, almost diamond-shaped. The apex of the process of the 9th sternite slightly cleft.

FEMALE GENITALIA.– Ostium with a sinuate double excavation. The antrum funnel-shaped, as long as wide. Apophyses anteriores short.

DISTRIBUTION.– The Mediterranean area, and known also from southern Russia.

BIOLOGY.– The moth flies from March to May and again in August and September. The hostplant is *Limoniastrum monopetalum* Boissier (Gibeaux, 1989b).

19. *Agdistis paralia* (Zeller, 1847)

Adactyla paralia Zeller, 1847: 899.   Pl. 3: 2. Figs 34, 171.

DIAGNOSIS.– Wingspan 21-25 mm. Colour pale grey, speckled with sparse brown scales. At the costa four spots and three other spots along the dorsal margin of the fold.

MALE GENITALIA.– Valvae bilobed. At apex of the first lobe with a small process both dorsally and ventrally. The costal arms long, slender, and pointed. Tegumen bilobed. The socii almost as long as the stalk of the uncus. The uncus rather small and rounded. The apex of the 9th sternite narrow and short.

FEMALE GENITALIA.– Ostium sinuate, doubly excavated; the ostium is not as wide as in *manicata*. The antrum twice as long as wide, obliquely ending, almost lying in the longitudinal direction of the body.

DISTRIBUTION.– Western Mediterranean area.

BIOLOGY.– The moth flies from April to September. The hostplants given by Bigot & Picard (1983a) are *Limonium vulgare* Miller and *L. virgatum* (Willd.) Fourr., without mentioning the life cycle.

20. *Agdistis bifurcatus* Agenjo, 1952

Agdistis bifurcatus Agenjo, 1952: 121.   Pl. 3: 3. Figs 35, 172.

DIAGNOSIS.– Wingspan 23-26 mm. Colour grey. At the costa four spots, and along the dorsal margin of the fold three spots.

MALE GENITALIA.– Valvae bilobed, at the distal margin of the first lobe a process half as long as the rather short distal lobe. The costal arms stout and slightly longer than the valve. Tegumen small, bilobed. The socii very long and wide. The uncus stoutly stalked, gradually widening. The apex of the 9th sternite deeply cleft.

FEMALE GENITALIA.– The ostium extended and slightly excavated in middle. Antrum laterally extended in the margin of the 7th tergite, and proximally of this funnel-shaped narrowing; as long as wide.

DISTRIBUTION.– Canary Islands, Spain and Morocco.

BIOLOGY.– The moth flies in March, May and June and again in September and October. The hostplant is unknown.

21. *Agdistis pseudocanariensis* Arenberger, 1973

Agdistis pseudocanariensis Arenberger, 1973a: 179.   Pl. 3: 4. Figs 36, 173.

DIAGNOSIS.– Wingspan 14-17 mm. Colour grey, speckled with grey-white scales. At the costa four evenly spaced spots, and along the dorsal margin of the fold three spots, the distal spot the biggest.

MALE GENITALIA.– Valvae gradually tapering towards the apex; with small saccular and cucullar processes. The costal arms well developed, cleft into two separate terminal processes. The 9th sternite bifid.

FEMALE GENITALIA.– Ostium centrally excavated. The antrum with a double sinuate distal margin, and proximally rounded. Apophyses anteriores small, as long as the papillae anales.

DISTRIBUTION.– Canary Islands, Morocco, southern Spain.

BIOLOGY.– The moth flies in January, March and April and again in November. The hostplant is unknown.

22. *Agdistis hartigi* Arenberger, 1973

Agdistis hartigi Arenberger, 1973b: 3.   Pl. 3: 5. Figs 37, 174.

DIAGNOSIS.– Wingspan 17 mm. Colour pale grey, within the fold grey-brown. At the costa four spots, the distance between the 2nd and 3rd spot the longest. Along the dorsal margin of the fold three spots, the shortest distance occurs between the 1st and the 2nd spots; the terminal spot is the largest.

MALE GENITALIA.– Valvae vesicular, with asymmetrical membraneous subterminal processes. The costal arms are asymmetrical, stout, and much shorter than the valvae.

FEMALE GENITALIA.– Ostium deeply excavated. Antrum almost squarish, the deeply excavated ostium extending as far as the proximal antrum margin. The 7th tergite is a small plate. The apophyses anteriores in form of short broad spines.

DISTRIBUTION.– Southern Spain, Sardinia.

BIOLOGY.– The moth flies in June and July. The hostplant is unknown.

23. *Agdistis betica* Arenberger, 1978

Agdistis betica Arenberger, 1978: 79.   Pl. 3: 6. Figs 38, 175.

DIAGNOSIS.– Wingspan 19-24 mm. Colour grey-brown. At the costa four spots, and along the dorsal margin of the fold three spots.

MALE GENITALIA.– The left valve with a membraneous process at two thirds from the base, and the right valve with a smaller process at the middle. The costal arms are asymmetrical.

FEMALE GENITALIA.– Ostium deeply excavating the antrum, which remains as narrow lateral margins. The 8th tergite deeply excavated toward the ostium. The 7th tergite small. A group of small spines in the junction between the ductus bursae and the ductus seminalis.

DISTRIBUTION.– Southern Spain.

BIOLOGY.– The moth flies in June and July. The hostplant is unknown.

REMARKS.– The species closely resembles *A. hartigi*, differing in the male genitalia in the shape of the valvae and costal arms; and in the female genitalia in the presence of a cluster of spines in the ductus bursae and seminalis.

SUBFAMILY: PTEROPHORINAE Zeller, 1841
= *PLATYPTILIINAE* Tutt, 1907.

The subfamily Pterophorinae is defined by the single cleft in the forewing.

Platyptilia Hübner, [1825]

Platyptilia Hübner, [1825]: 429.
   Type species: *Alucita megadactyla* [Denis & Scheffermüller], 1775 (=*Alucita gonodactyla* [Denis & Schiffermüller], 1775); subsequent designation (Tutt, 1905).
Platyptilus Zeller, 1841; emendation.
Fredericina Tutt, 1905: 37.
   Type species: *Alucita calodatyla* [Denis & Schiffermüller], 1775; original designation.

DIAGNOSIS.– Costal triangle on forewing in most species well developed. Forewing vein R1 present. Third lobe of hindwing with centrally placed scale-tooth.

MALE GENITALIA.– Valvae symmetrical, lanceolate. Sacculus simple, almost as long as the valve. Cucullus poorly developed. Anellus arched. Tegumen arched with a well developed uncus. Saccus in shape of a pentagular plate with variable distal dentation. Aedeagus curved, provided with a cornutus in form of minute spiculae.

FEMALE GENITALIA.– Antrum in the shape of a tubular sclerotised segment of variable length, ending centrally at the distal margin of the 8th sternite. Ductus bursae without a sclerotised segment. Lamina postvaginalis with two blotches beside the ostium bursae, laterally extending into the short apophyses anteriores. Signum double, horn-like.

DISTRIBUTION.– Holartic, South America, Afrotropical and Indo-Australian regions.

BIOLOGY.– Asteraceae. (Hofmann, 1896; Barnes & Lindsey, 1921; Schwarz, 1953; Gielis, bred).

24. *Platyptilia tesseradactyla* (Linnaeus, 1761)

Phalaena Alucita tesseradactyla Linnaeus, 1761: 370.　　　　Pl. 3: 7. Figs 39, 176.
Pterophorus fischeri Zeller, 1841: 781.

DIAGNOSIS.– Wingspan 17-20 mm. Colour brown-grey. At the top of the costal triangle a small black line, outlining the terminal part of the triangle.

MALE GENITALIA.– Valvae terminally distinctly tapered. Anellus arms rather stout, short, not cleft.

FEMALE GENITALIA.– Sclerotised segment of antrum one and a half times as long as wide. The blotches of the lamina postvaginales large, and almost as long as the antrum.

DISTRIBUTION.– Northern and Central Europe, extending eastwards into Russia, in Norway as far north as 70° latitude. Also known from North America.

BIOLOGY.– The moth flies in June and July. The species prefers mountainous areas, with pine-forests and a sandy soil. The hostplants are *Helichrysum arenarium* L. (Hofmann, 1896; Hannemann, 1977b) and *Antennaria dioica* L. (Gartner, 1862; Hofmann, 1896; Huggins, 1939; Beirne, 1954; Gozmány, 1962; Hannemann, 1977b; Buszko, 1986). The larva feeds initially in the soft central core of the stem of the plant and hibernates there. In spring the young shoots are spun together and the top parts are eaten, causing stunting of the plant. Pupation in the spinning. The pupal period is approximately 3 weeks.

25. *Platyptilia farfarellus* Zeller, 1867

Platyptilus farfarellus Zeller, 1867a: 334.　　　　Pl. 3: 8. Figs 40, 177.

DIAGNOSIS.– Wingspan 18-19 mm. Colour reddish brown, markings dark brown. A white sub-terminal line in both forewing lobes. The hind tibiae and the first tarsal segment pale ochreous brown, in contrast to *P. gonodactyla* in which they are white.

MALE GENITALIA.– Valvae gradually tapered toward tip. Uncus basally wide and gradually tapering toward apex. Anellus arms rather slender, with a small hook at middle.

FEMALE GENITALIA.– Antrum gradually narrowing, sclerotised and three times as long as the ductus bursae. Apophyses anteriores with three tips.

DISTRIBUTION.– Central and southern Europe and Asia Minor. Also recorded from Japan.

BIOLOGY.– The moth flies in May and again in July to September. The hostplants are *Senecio vernalis* W. & K. (Hofmann, 1896; Mitterberger, 1912; Gozmány, 1962; Hannemann, 1977b; Burmann, 1986) and *S. viscosus* L. (Hofmann, 1896; Mitterberger, 1912; Gozmány, 1962; Burmann, 1986; Arenberger & Jaksic, 1991), growing in sandy localities and along edges of woods. The spring larvae spin the central shoots together and feed on the root-stock, excavating this. Pupation in the excavated space. The summer brood lives on the flower-buds, and pupates in the flower-shoot.

REMARKS.– Yano (1963) mentions the species from Japan, and gives *Erigeron canadensis* L. and *Calendula arvensis* L. as hostplants. Also other hostplants as recorded. These may be regarded as possible for Europe, but need verification.

26. Platyptilia nemoralis Zeller, 1841

Platyptilia nemoralis Zeller, 1841: 778.　　　　　　　　　Pl. 3: 9. Figs 41, 178.
Platyptilia nemoralis var. *saracenica* Wocke [in: Staudinger & Wocke], 1871, nr. 3127a.
Platyptilia grafii Zeller, 1873b: 139.

DIAGNOSIS.– Wingspan 27-32 mm. Colour pale brown, markings dark brown. Forewing with wide terminal parts, termen of first lobe gradually curved into the apex. Scale-tooth at the dorsum of the third lobe of the hindwing well developed.

MALE GENITALIA.– Cucullus and sacculus margins almost parallel. Uncus of moderate size. Saccus slightly excavated. Anellus arms wide, bidentate.

FEMALE GENITALIA.– Antrum long, sclerotised, five times as long as the unsclerotised part of ductus bursae. The blotches at the lamina postvaginalis extended. Apophyses posteriores long.

DISTRIBUTION.– Western, southern and Central Europe.

BIOLOGY.– The moth flies in July and August. In southern areas an early spring generation is stated to be on the wing in March by LHomme (1939). The hostplants, *Senecio fuchsii* Gmelin (Hannemann, 1977b), *S. fluviatilis* Wallr. (Hannemann, 1977b), *S. sarracenicus* L. (Hofmann, 1896; Mitterberger, 1912; Gozmány, 1962) and *S. nemorensis* L. (Hofmann, 1896; Mitterberger, 1912; Gozmány, 1962; Hannemann, 1977b; Arenberger & Jaksic, 1991), grow in wet places in woodlands, in marshes and along brooks. The larva feeds in the shoots, often several larvae in each shoot, which becomes thickened. Faecal grains are removed through a small hole in the stem. Pupation in the feeding chambers, the pupal period lasting for 14 days.

27. Platyptilia gonodactyla ([Denis & Schiffermüller], 1775)

Alucita gonodactyla [Denis & Schiffermüller], 1775: 320.　　　Pl. 3: 10; pl. 15: 3. Figs 42, 179.
Alucita megadactyla [Denis & Schiffermüller], 1775: 145.
Alucita diptera Sulzer, 1776: 163.
Alucita trigonodactyla Haworth, 1811: 478.
Platyptilia farfara Gregson, 1885: 151.

DIAGNOSIS.– Wingspan 22-28 mm. Colour pale brown, mixed with grey scales. Markings dark brown. The termen of the first lobe of the forewing is sinuate and ends into an acute apex.

MALE GENITALIA.– Valvae gradually, but only slightly, tapering. Uncus of moderate size. Saccus deeply excavated. Anellus arms slender, with a large spine at middle.

FEMALE GENITALIA.– Antrum hardly narrowing; the sclerotised part seven times as long as the membraneous part of the ductus bursae.

DISTRIBUTION.– Europe, except for the most southerly parts, through Russia and Asia Minor into India.

BIOLOGY.– The moth flies from May to October in two generations. The hostplants, *Tussilago farfara* L. (Hofmann, 1896; Gozmány, 1962; Emmet, 1979; Gielis, bred) and *Petasites* sp. (Hofmann, 1896; Beirne, 1954; Hannemann, 1977b; Arenberger & Jak-

40

sic, 1991; Huemer, pers. comm.), grow in relatively wet places in woodlands, along brooks and beside ponds, on roadsides, arable land, etc. The larvae of the spring generation start to feed on the leaves, overwinter in the rootstock, eating their way up inside the flowershoots and finally feed on the bottom parts of the flowers. Pupation in the flower remains or on the underside of a leaf. The summer generation larvae start as leaf miners, later feed on the underside of the leaves, where they also pupate.

REMARKS.– The synonymy of *gonodatyla* and *megadactyla* was discussed in detail by Karsholt & Gielis (1995).

28. *Platyptilia calodactyla* ([Denis & Schiffermüller], 1775)

Alucita calodactyla [Denis & Schiffermüller], 1775: 146.   Pl. 4: 1, 2. Figs 43, 180.
Alucita petradactyla Hübner, [1819]: T. 7, figs 37, 38.
Pterophorus zetterstedtii Zeller, 1841: 777.
Platyptilia taeniadactyla South, 1882: 34.
Platyptilia leucorrhyncha Meyrick, 1902: 217.
Platyptilia calodactyla var. *doronicella* Fuchs, 1902: 329.

DIAGNOSIS.– Wingspan 18-25 mm. Colour yellow to orange brown. Termen of forewing lobes almost straight.

MALE GENITALIA.– Valvae gradually tapering. Uncus rather stout. Saccus slightly excavated, basally widely lobed. Anellus arms slender with a conspicuous median spine.

FEMALE GENITALIA.– Antrum gradually tapering, six times as long as the membraneous part of the ductus bursae.

DISTRIBUTION.– Europe, except for the far northern and southern parts of the area. In Norway as far north as 70º 20' latitude.

BIOLOGY.– The moth flies from June to August. The moth is more common in mountainous areas. The hostplants are *Solidago virgaurea* L. (Beirne, 1954; Gozmány, 1962), *Senecio nemorensis* L. (Mitterberger, 1912), *S. sylvaticus* L., *Doronicum* sp. (Arenberger & Jaksic, 1991) and, in Central Norway, *Erigeron* acer ssp. *politus* Fries. (Karsholt, bred). The larva feeds in the stems, near the ground, and pupates in the stem.

29. *Platyptilia iberica* Rebel, 1935

Platyptilia iberica Rebel, 1935: 10.   Pl. 4: 3. Figs 44, 181.
Platyptilia iberica nevadensis Rebel, 1935: 10.

DIAGNOSIS.– Wingspan 18-23 mm. Colour yellow to grey brown. Forewing more slender than in *P. gonodactyla* and *P. calodactyla*. The costal triangle ill-developed; two pale transverse fasciae present on each lobe.

MALE GENITALIA.– Valvae gradually narrowing. Uncus slender. Saccus with rounded edges. Anellus arms strong, wide, with two well developed spines.

FEMALE GENITALIA.– Not significantly different from *P. gonodactyla*.

DISTRIBUTION.– Mountains of west-central and southern parts of Spain, France (Pyrénées).

BIOLOGY.– The moth flies in July. The hostplant is unknown, but the author collected the species around *Senecio* sp. amongst which the imagines were hidden through the day, near Hoyos del Espino, Prov. Avila.

30. *Platyptilia isodactylus* (Zeller, 1852)

Pterophorus isodactylus Zeller, 1852: 328.   Pl. 4: 4. Figs 45, 182.
Platyptilia brunneodactyla D. Lucas, 1955: 255.

DIAGNOSIS.– Wingspan 19-29 mm. Colour brown. Forewing rather slender, as in *P. iberica*. The costal triangle reduced to a few small longitudinal spots.

MALE GENITALIA.– Valvae rounded. Uncus stout. Saccus in shape of a pentagram. Anellus arms short, with a spine at middle.

FEMALE GENITALIA.– Sclerotised part of the antrum about twice as long as the membraneous part of the ductus bursae, and gradually narrowing.

DISTRIBUTION.– Central and western Europe, extending to the south as far as to Spain and Morocco. Also recorded from Japan.

BIOLOGY.– The moth flies from June to September. The hostplants are *Senecio* spp. *S. aquaticus* Hudson (South, 1881; Mitterberger, 1912; Beirne, 1954; Gozmány, 1962; Hannemann, 1977b; Emmet, 1979), *S. nemorensis* L. (de Graaf, 1868) and *S. jacobeae* L. (Gielis, bred) are verified host records. The larva feeds in the spun top-shoots, later infesting the stem. A small opening is used to drop the faecal remains, which tend to remain near the opening because they are enveloped by spinnings. Pupation in the stem.

Gillmeria Tutt, 1905

Gillmeria Tutt, 1905: 37.
   Type species: *Alucita ochrodactyla* [Denis & Schiffermüller], 1775; original designation.

DIAGNOSIS.– The head with a frontal scale brush. Palpi long and slender, porrect. Forewing in general rather acutely pointed. Forewing markings little developed, the costal triangle is only indicated by some small lines and dots. At the dorsum of the third lobe of the hindwing a weakly developed scale-tooth.

MALE GENITALIA.– Valvae symmetrical. Basal section of sacculus often wider than, and well differentiated from the distal section.

FEMALE GENITALIA.– Closely resembling *Platyptilia* Hübner.

DISTRIBUTION.– The genus has a Holarctic distribution, and a single specimen of *G. pallidactyla* has been collected in Brazil.

BIOLOGY.– The known hostplants belong to the Asteraceae genera: *Achillea* and *Tanacetum*.

31. *Gillmeria miantodactylus* (Zeller, 1841)

Pterophorus miantodactylus Zeller, 1841: 767.   Pl. 4: 5. Figs 46, 183.

DIAGNOSIS.– Wingspan 17-22 mm. Colour grey to yellow brown. The markings consist of small ochreous costal streaks.

MALE GENITALIA.– Valvae rounded, with a basally wide sacculus which extends as a very narrow sclerotised marginal ridge. Saccus small, roundish. Anellus arms simple and short.

FEMALE GENITALIA.– Antrum rather wide, and sclerotised over a segment three times as long as the membraneous part of the ductus bursae.

DISTRIBUTION.– South-eastern Europe, extenting into Asia Minor. A single specimen has been recorded from the Pyrenees in France.

BIOLOGY.– The moth flies in July. The presumed hostplant is *Scabiosa ochroleuca* L. (Rothschild, 1913), but no breeding records have been published yet.

32. *Gillmeria pallidactyla* (Haworth, 1811)

Alucita pallidactyla Haworth, 1811: 478.   Pl. 4: 6. Figs 11, 47, 184.
Pterophorus migadactylus Curtis, 1827: 161.
Alucita ochrodactyla Treitschke, 1833: 225.
Pterophorus marginidactylus Fitch, 1854: 848.
Pterophorus nebulaedactylus Fitch, 1854: 849.
Platyptilus bertrami Roessler, 1864: 54.
Platyptilus bischoffi Zeller, 1867a: 333.
Pterophorus cervinidactylus Packard, 1873: 266.
Platyptilus adustus Walsingham, 1880: 5.

DIAGNOSIS.– Wingspan 23-27 mm. Palpi slender, twice as long as diameter of eye. Frons conically protruding, about one and a half times diameter of eye. Forewing colour brown-yellow, markings pale brown. The markings consist of an indistinct costal triangular spot and costal darkening between the wingbase and the triangle. Scale tooth on the dorsum of the third lobe of the hindwing faintly indicated. The hindlegs brown-grey, distally gradually paler brown, thus differing from the similar *G. tetradactyla* which has ringed hindlegs.

VARIATION.– Small and greyish tinged specimens occur in colder areas of its range.

MALE GENITALIA.– Valvae symmetrical, rounded. Sacculus basally well developed, not reaching apex of valve. Uncus stout. Anellus arms stout, blunt-ending, with a lateral pointed projection.

FEMALE GENITALIA.– Antrum elongate tube-like, four times as long as wide. Ductus bursae a little longer than antrum.

DISTRIBUTION.– Holarctic region, and a single record from Brazil.

BIOLOGY.– The moth flies in June and July. The hostplants, *Achillea ptarmica* L. (Hofmann, 1896; Gozmány, 1962; Hannemann, 1977b) and *A. millefolium* L. (Hofmann, 1896; Beirne, 1954; Gozmány, 1962; Hannemann, 1977b; Emmet, 1979), grow on

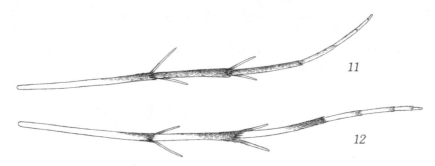

Figs 11, 12. Hind tibia and tarsus of 11, *Gillmeria pallidactyla* and
12, *G. tetradactyla*. (R. Nielsen del.).

road sides, edges of woodland and sandy meadows. The young larva spins the central
shoots together and feeds on the leaves, later it bores a hole in the stem and excavates
this. The faecal remains are removed through a small hole in the stem. Pupation on
the underside of a leaf.

33. *Gillmeria tetradactyla* (Linnaeus, 1758)

Phalaena Alucita tetradactyla Linnaeus, 1758: 542.   Pl. 4: 7. Figs 12, 48, 185.
Alucita ochrodactyla [Denis & Schiffermüller], 1775: 145.
Platyptilus dichrodactylus Mühlig, 1863: 213.
Platyptilia ochrodactyla var. *borgmanni* Roessler, 1881: 220.
Platyptilia ochrodactyla var. *bosniaca* Rebel, 1904: 323.

DIAGNOSIS.– Wingspan 24-28 mm. Colour reddish yellow. Markings pale brown as in *G.
pallidactyla*, but more well developed. Frons and palpi as in *G. pallidactyla*. Hindlegs
ringed, grey-brown and dark brown, see illustration.

MALE GENITALIA.– Valvae lanceolate, acute at apex. Uncus stout. Saccus small. Anellus
arms simple.

FEMALE GENITALIA.– Ostium ring-shaped, large. Ductus bursae membraneous.

DISTRIBUTION.– Europe.

Biology.– The moth flies from June to August. The hostplant, *Tanacetum vulgare* L.
(Hofmann, 1896; Beirne, 1954; Gozmány, 1962; Hannemann, 1977b; Emmet, 1979;
Arenberger & Jaksic, 1991), grows along roads, in woodlands and along brooks and
ponds. The larva feeds in the central shoot. Pupation on the stem or under a leaf.

Lantanophaga Zimmerman, 1958

Lantanophaga Zimmerman, 1958: 400-402.
   Type species: *Oxyptilus pusillidactylus* Walker, 1864; original designation.

DIAGNOSIS.– Third lobe of hindwing with distally placed scale-tooth. Costal triangle well
developed. Forewing vein R1 present. First tarsal segment of hind leg half as long as
the combined length of segments two to five.

MALE GENITALIA.– Valvae resembling a bird-head, with extension of the top. Sacculus bilobed, the proximal and distal halves of equal length. A long and slender, forked saccus. Tegumen simple. Uncus slender.

FEMALE GENITALIA.– As in the genus *Platyptilia*, but a sclerotised plate in the ductus bursae. The lamina postvaginalis well developed, showing a bilobed sclerotised plate distal to the ostium, and laterally progressing into the apophyses anteriores. A pair of S-shaped signa.

DISTRIBUTION.– Tropical and subtropical regions of the world.

BIOLOGY.– *Lantana* (Fletcher, 1921; Amsel, 1955a).

34. *Lantanophaga pusillidactylus* (Walker, 1864)

Oxyptilus pusillidactylus Walker, 1864: 933.   Pl. 4: 8. Figs 49, 186.
Platyptilia technidion Zeller, 1877: 13.
Platyptilia hemimetra Meyrick, 1886: 18.
Platyptilia lantana Busck, 1914: 103.
Platyptilia lantanadactyla Amsel, 1951a: 66.

DIAGNOSIS.– Wingspan 11-14 mm. Colour ochreous brown. Eight small white dots along costa between base and the base of cleft. An indistinct costal triangle (dark brown), with small free margin to base of cleft. Hindwing light orange-yellow. Irregularly distributed black scales on hind margin of third lobe. In distal part a scale-tooth, close to the apex.

The species resembles *Stenoptilodes taprobanes* very much, but differs in the darker colour, the generally smaller size and the length of the first tarsal segment in the hind leg.

MALE GENITALIA.– See genus description.

FEMALE GENITALIA.– See genus description.

DISTRIBUTION.– Tropical and subtropical areas of the world, reaching its northern limit of distribution in Morocco, Madeira and the Canary Islands.

BIOLOGY.– The moth flies in July and September to December. The hostplant is *Lantana camara* L. (Amsel, 1951a). Eggs are deposited on the flower-bracts. The larva feeds on the flowers and pupates thereon.

Stenoptilodes Zimmerman, 1958

Stenoptilodes Zimmerman, 1958: 407.
   Type species: *Platyptilus littoralis* Butler, 1882; original designation.

DIAGNOSIS.– Costal triangle well developed. Forewing vein R1 present. Third lobe of hindwing with distally placed scale-tooth. The first tarsal segment in the hind leg is longer than half the combined length of the other tarsal segments.

MALE GENITALIA.– Valvae with bird-head-like apex. Sacculus bilobed, the apical section small. Vinculum arched, saccus not developed. Tegumen simple. Uncus long and slender.

FEMALE GENITALIA.– Antrum squarish to rectangular, sclerotised, centrally placed at the distal margin of the 8th sternite. A sclerotised plate in the ductus bursae. The lamina postvaginalis not developed. Apophyses anteriores short. Two horn-like signa.

DISTRIBUTION.– Tropical and subtropical regions of the world.

BIOLOGY.– Depending on the faunal area, a great number of hostplants have been recorded, belonging to several families. See *taprobanes*.

35. *Stenoptilodes taprobanes* (Felder & Rogenhofer, 1875)

Amblyptilia taprobanes Felder & Rogenhofer, 1875: plate 140.   Pl. 4: 9. Figs 50, 187.
Platyptilia brachymorpha Meyrick, 1888: 240.
Platyptilia seeboldi Hofmann, 1898b: 33.
Platyptilia terlizzii Turati, 1926: 67.
Amblyptilia zavatterii Hartig, 1953: 67.
Platyptilia legrandi Bigot, 1962: 86.

DIAGNOSIS.– Wingspan 12-17 mm. Colour grey-brown. See genus description.

MALE GENITALIA.– See genus description.

FEMALE GENITALIA.– See genus description.

DISTRIBUTION.– Mediterranean area. The species is recorded from the entire tropical and subtropical zone.

BIOLOGY.– In our area the moth flies from March to October. The hostplant is *Spergularia media* L. (=*S. marginata* Kitt) (Nel & Prola, 1989). This plant grows in saltmarshes along the coasts of the Mediterranean area. The larva feeds on the flowers, and eats out a shelter in a stem or the flower base. The hostplants recorded outside Europe are *Celsia coromandeliana* Vahl (Fletcher, 1921), *Antirrhinum* sp., *Clinopodium vulgare* L., *Vaccinium* sp., *Campylanthus salsoloides* Roth, *Centipeda minima* O. Kuntze (Yano, 1963) and *Plectranthus* sp. On these plants the larva feeds on the flower-buds and flowers. Gaj (1959) mentions feeding on the fruit of *Limnophila* and *Veronica* and the unripe seeds of *Pentstemon*. Pupation on the hostplant. The pupal period lasts seven to eight days in India (Fletcher, 1921).

Paraplatyptilia Bigot & Picard, 1986

Mariana Tutt, 1907: 160; homonym of *Mariana* Locard, 1899 (Mollusca).
   Type species: *Pterophorus metzneri* Zeller, 1841; monotypy.
Paraplatyptilia Bigot & Picard, 1986: [17]; replacement name for *Mariana* Tutt, 1907.

DIAGNOSIS.– Costal triangle well-developed. Forewing vein R1 present. Third lobe of hindwing with distally placed scale-tooth.

MALE GENITALIA.– Valvae with an apex resembling a bird-head. Sacculus bilobed, the apical section small. The vinculum has a small saccus. Tegumen bilobed. Uncus spoon-like.

FEMALE GENITALIA.– Antrum tube-like, localized laterally in the sclerotised end plate of the

8th sternite. In the ductus bursae a well developed sclerotised plate. The lamina postvaginalis centrally fused with the sclerotised distal margin of the 8th sternite, and laterally extending into the apophyses anteriores. Two horn-like signa.

DISTRIBUTION.– Holarctic and Neotropical regions.

BIOLOGY.– Probably on Fabaceae and Scrophulariaceae (LHomme, 1939).

36. *Paraplatyptilia metzneri* (Zeller, 1841)

Pterophorus metzneri Zeller, 1841: 783.   Pl. 4: 10. Figs 51, 188.
Pterophorus bollii Frey, 1856: 403.

DIAGNOSIS.– Wingspan 23-27 mm. Colour grey with numerous brown scales. See genus description.

MALE GENITALIA.– See genus description.

FEMALE GENITALIA.– See genus description.

DISTRIBUTION.– Alpine region in Central and South-East Europe, probably extending into Russia.

BIOLOGY.– The moth flies from June to August. It occurs on mountain slopes and passes at altitudes above 1200 m. The presumed hostplant is an *Astragalus* sp. (Arenberger & Jaksic, 1991).

REMARKS.– The species belonging to this genus are closely related and resemble each other very much. The distribution of the present species in Asia Minor is possible, but needs confirmation.

Amblyptilia Hübner, [1825]

Amblyptilia Hübner, [1825]: 430.
     Type species: *Alucita acanthadactyla* Hübner, [1813]: tab. 5, figs 23, 24; subsequent designation (Tutt, 1905).
Amplyptilia Hübner, [1825]; incorrect (of multiple original) spelling.
Amblyptilus Wallengren, 1862; emendation.

DIAGNOSIS.– Costal triangle well developed. Forewing vein R1 present. Third lobe of hindwing with centrally placed scale-tooth.

MALE GENITALIA.– Apex of valvae in shape of a bird-head. Sacculus not lobed (in *A. punctidactyla* very narrow in central part). Basally the vinculum has a bristle-like saccus. Tegumen bilobed. Uncus simple.

FEMALE GENITALIA.– The antrum tube-like, localized laterally in the heavily sclerotised terminal plate of the 8th sternite. In the ductus bursae there is a small sclerotised plate. The lamina postvaginalis is fused with the sclerotised distal margin of the 7th sternite and extends laterally into the apophyses anteriores. Signum double, horn-like.

DISTRIBUTION.– Holarctic, Afrotropical, Indo-Australian and Neotropical regions.

BIOLOGY.– Polyphagous: among the hostplants are representatives of the families Scrophulariaceae, Lamiaceae, Geraniaceae and Fabaceae (Frey, 1856; Hofmann, 1896; Tutt, 1896; Gielis, bred).

37. *Amblyptilia acanthadactyla* (Hübner, [1813])

Alucita acanthadactyla Hübner, [1813]: t. 5, figs 23, 24. Pl. 5: 1; pl. 15: 4. Figs 52, 189.
Amblyptilia calaminthae Frey, 1886: 16.
Amblyptilia tetralicella Hering, 1891 (Hofmann, 1896); *nomen nudum*.

DIAGNOSIS.– Wingspan 17-23 mm. Colour dark red-brown, speckled with white scales.

MALE GENITALIA.– See genus description. Differs from *punctidactyla* by the indentated sacculus and the straighter anellus arms.

FEMALE GENITALIA.– Ostium distinctly incurved. Lamina postvaginalis narrower than in *punctidactyla*.

DISTRIBUTION.– Europe, North Africa and Asia Minor.

BIOLOGY.– The moth flies in two generations. The spring generation flies in June to August. The summer generation starts to fly in October, and may be seen through the winter in mild conditions, until April and May. The species occurs in a great number of environments. The larva is polyphagous and is recorded from a great number of hostplants: *Stachys sylvatica* L.(Frey, 1870; Mitterberger, 1912; Emmet, 1979; Nel, 1989b), *Ononis natrix* L. (Gielis, bred), *O. repens* L. (Mitterberger, 1912; Hannemann, 1977b;), *O. spinosa* L. (Hofmann, 1896; Mitterberger, 1912;Hannemann, 1977b), *Mentha* sp. (Mitterberger, 1912; Hannemann, 1977b), *Salvia* sp. (Mitterberger, 1912; Hannemann, 1977b), *Euphrasia* sp. (Mitterberger, 1912; Hannemann, 1977b), *Teucrium scorodonia* L. (Hannemann, 1977b; Nel, 1989b), *Chenopodium* sp. (Hannemann, 1977b), *Calluna vulgaris* L. (South, 1881; Mitterberger, 1912; Hannemann, 1977b), *Erica tetralix* L. (Hofmann, 1896; Buhl et al., 1983), *Geranium robertianum* L. (Hannemann, 1977b), *Bartsia* sp. (Hannemann, 1977b), *Carlina* sp. (Hannemann, 1977b), *Lavandula stoechas* L. (Nel, 1989b), *Vaccinium oxycoccus* L. (Mitterberger, 1912; Hannemann, 1977b), *Calamintha nepeta* (L.) Savi (Frey, 1886; Hofmann, 1896; Mitterberger, 1912; Hannemann, 1977b; Nel, 1989b), *Jurinea* sp. (Mitterberger, 1912; Hannemann, 1977b) and *Nepeta* sp. (Emmet, 1979; Hannemann, 1977b; Arenberger & Jaksic, 1991; Gielis, bred). The larva feeds on the young leaves and flowers. Pupation occurs on the hostplant.

38. *Amblyptilia punctidactyla* (Haworth, 1811)

Alucita punctidactyla Haworth, 1811: 479. Pl. 5: 2. Figs 53, 190.
Alucita cosmodactyla Hübner, [1819]: t. 7, figs 35, 36.
Alucita ulodactyla Zetterstedt, 1840: 1012.
Platyptilus cosmodactylus var. *stachydalis* Frey, 1870: 290.

DIAGNOSIS.– Wingspan 18-23 mm. Colour brown-grey, darker than the previous species, and more speckled with white scales.

MALE GENITALIA.– Sacculus indentated. Anellus arms curved and more stoutly built than in *A. acanthodactyla*.

FEMALE GENITALIA.– Ostium of the antrum more flattened, obliquely ending. Lamina postvaginalis wider compared to length than in previous species.

DISTRIBUTION.– England, western Europe (except for the Netherlands), Central Europe and the southern half of Scandinavia.

BIOLOGY.– The first generation of the moth flies in June and July; the second generation in September, and after the winter, again in April and May. Like the previous, this is a polyphagous species. Recorded hostplants are: *Stachys sylvatica* L. (Frey, 1870; Hannemann, 1977b; Emmet, 1979; Buszko, 1986), *Aquilegia vulgaris* L. (Leech, 1886; Mitterberger, 1912; Hannemann, 1977b; Nel, 1989b), *Geranium pratense* L. (Leech, 1886; Hannemann, 1977b; Nel, 1989b), *Erodium cicutarium* L. (Mitterberger, 1912; Hannemann, 1977b), *Salvia glutinosa* L. (Mitterberger, 1912; Hannemann, 1977b; Emmet, 1979; Buszko, 1986; Arenberger & Jaksic, 1991), *Primula* sp., *Ononis spinosa* L. (Cansdale, 1955) and *Prunella* sp. (Gielis, bred). The larva feeds on the flowers and seeds. Pupation on the hostplant.

Stenoptilia Hübner, [1825]

Stenoptilia Hübner, [1825]: 430.
  Type species: *Phalaena Alucita pterodactyla* Linneaus, 1761; subsequent designation (Tutt, 1905).
Mimaeseoptilus Wallengren, 1862: 18.
  Type species: *Alucita mictodactyla* [Denis & Schiffermüller], 1775; subsequent designation (Meyrick, 1910).
Mimeseoptilus Zeller, 1867; emendation.
Mimaesoptilus Snellen, 1884; incorrect spelling.
Doxosteres Meyrick, 1886: 10.
  Type species: *Pterophorus canalis* Walker, 1864; monotypy.
Mimaesioptilus Barrett, 1904; incorrect spelling.
Adkinia Tutt, 1905: 37.
  Type species: *Phalaena bipunctidactyla* Scopoli, 1763; original designation.
Adkina Yano, 1963; incorrect spelling.

DIAGNOSIS.– Costal triangle reduced to some dark spots. Forewing vein R1 present. Third lobe of hindwing without a scale-tooth.

MALE GENITALIA.– Valvae resembling a bird-head. Sacculus bilobed, apical section small. Vinculum arched, saccus not developed. Tegumen simple. Uncus small.

FEMALE GENITALIA.– Antrum tube-like, centrally placed at the distal margin of the 8th sternite. In ductus bursae a sclerotised plate. The lamina postvaginalis not developed, and also the apophyses anteriores absent. In bursa copulatrix a double horn-like signum.

DISTRIBUTION.– Holarctic, Neotropical, Afrotropical and Indo-Australian regions.

BIOLOGY.– Larvae on Gentianaceae, Scrophulariaceae, Saxifragaceae, Primulaceae and Dipsacaceae (Nel, 1986d).

REMARKS.– French authors have recently described a number of new species of *Stenoptilia*. They assumed that larvae living on a different hostplant, even as very closely related species of hostplant, must belong to a different species of moth. To underline this assumption they used the setal pattern of the larvae. The larvae used to illustrate this idea were of the fifth to the seventh larval instars. Already Wasserthal (1970) indicated this to be incorrect because of the adaptation of the larvae to their host-plants. The specific setal pattern varies in the later larval stages, and shows a relation to the host plant. Therefore only the egg-larvae or the first instar larvae can be used in this respect. The species described by these French authors are therefore synonymy-sed with already described species with distinctive genital structures.

39. *Stenoptilia graphodactyla* (Treitschke, 1833)

Alucita graphodactyla Treitschke, 1833: 233.   Pl. 5: 3. Figs 54, 191.

DIAGNOSIS.– Wingspan 18-25 mm. Colour grey-brown, markings black-brown. Forewing with two wedge-like spots before the base of the cleft. Second lobe of forewing with two black longitudinal lines, and one larger black elongated spot in the first lobe. Both forewing lobes have a black, continuous, basal fringe line along the termen.

MALE GENITALIA.– Valvae rather stout. Anellus arms two thirds as long as the tegumen. Uncus small, and the apex is at the margin of the tegumen.

FEMALE GENITALIA.– Ostium wide, slightly excavated, sharply narrowing, funnel-shaped and progressing into the gradually narrowing antrum. The width of the ostium and the length of the ostium and antrum are equal. Ductus bursae twice as long as antrum, stout, with a long sclerotised plate.

DISTRIBUTION.– Central and South-East Europe.

BIOLOGY.– The moth flies in July and August. The hostplants belong to the genus *Genti-ana*. Recorded are *G. lutea* L. (Gozmány, 1962; Hannemann, 1977b), *G. verna* L. (Hannemann, 1977b), *G. clusii* Perr. & Song. (Huemer, bred), *G. asclepiadea* L. (Mit-terberger, 1912; Osthelder, 1939; Gozmány, 1962; Hannemann, 1977b; Buszko, 1986; Gibeaux, 1989c) and *G. pneumonanthe* L. (Hannemann, 1977b; Arenberger & Jak-sic, 1991). The larva feeds on the leaves and flowers which are spun together, later on the seedheads. Pupation on the stem or a neighbouring plant.

REMARKS.– The hostplant records of *G. lutea* may refer to *S. lutescens* Herrich-Schäffer, 1855; the record of *G. pneumonanthe* is likely to refer to the next species: *S. pneumo-nanthes* Büttner.

40. *Stenoptilia pneumonanthes* (Büttner, 1880)

Mimeseoptilus pneumonanthes Büttner, 1880: 472.   Pl. 5: 4. Figs 55, 192.
Stenoptilia nelorum Gibeaux, 1989c: [13].
Stenoptilia arenbergeri Gibeaux, 1990b: 220.

DIAGNOSIS.– Wingspan 17-22 mm. Colour grey-brown. Forewing with a small double spot at the base of the cleft. In the first forewing lobe a longitudinal black spot and in the second lobe only a small group of black scales, not in shape of a well-defined spot. The fringes of the termen of both forewing lobes have a continuous black basal line.

MALE GENITALIA.– Resembling the previous species very closely.

FEMALE GENITALIA.– Closely resembling the previous species. The ostium seems to be deeper excavated than in *S. graphodactyla*. However in specimens from the Netherlands a flatter ostium is seen.

DISTRIBUTION.– Northern parts of Central and western Europe, and in the southern half of Scandinavia.

BIOLOGY.– The moth flies from the end of June to mid-September. The hostplants are *Gentiana pneumonanthe* L. (Beirne, 1954; Gozmány, 1962; Hannemann, 1977b; Emmet, 1979; Buszko, 1986; Gibeaux, 1989c; Gielis, bred) and *G. cruciata* L. (Gibeaux, 1989c; Gielis, bred). The larva feeds on the flowers and young seed capsules. Pupation on the plant. Pupal period approximately 14 days.

REMARKS.– The genitalia illustrated by Gibeaux in the description of *S. arenbergeri* are identical to those of *S. pneumonanthes*. Specimens identified by Gibeaux as *S. arenbergeri* and *S. pneumonanthes* showed no significant differences.

41. *Stenoptilia gratiolae* Gibeaux & Nel, 1990

Stenoptilia gratiolae Gibeaux & Nel, 1990b: 200.   Pl. 5: 5. Figs 56, 193.
Stenoptilia paludicola, auct., nec Wallengren, 1862: 18.

DIAGNOSIS.– Wingspan 18-24 mm. Colour dark red-brown. Forewings with a fused double spot at the base of the cleft. The dark spots in the lobes absent, only some isolated dark scales are present. Fringes of the termen of both lobes with a continuous black basal line.

MALE GENITALIA.– Valvae rather slender. Anellus arms two thirds as long as tegumen. Uncus extending for half its length beyond the margin of the tegumen.

FEMALE GENITALIA.– Ostium slightly excavated. The antrum is slightly obliquely arranged, gradually narrowing, about as long as wide.

DISTRIBUTION.– Central and western Europe.

BIOLOGY.– The moth flies in May and later in July and August. The hostplant is *Gratiola officinalis* L. (Buszko, 1986; Arenberger, 1990c; Gibeaux & Nel, 1990b), which grows on wet meadows and marshes. This plant is rarely recorded, and the sparse collecting data are probably caused by the scarcity of the hostplant. The larva feeds on the flowers and the seedheads. Pupation occurs along the stem.

42. *Stenoptilia pterodactyla* (Linnaeus, 1761)

Alucita pterodactyla Linnaeus, 1761: 370.   Pl. 5: 6. Figs 57, 194.
Pterophorus fuscus Retzius, 1783: 35.
Alucita fuscodactyla Haworth, 1811: 476.
Alucita ptilodactyla Hübner, [1813]: T. 3, fig. 16.
Mimaeseoptilus paludicola Wallengren, 1862: 18.

DIAGNOSIS.– Wingspan 20-26 mm. Colour pale red-brown to yellow-brown. Dorsal margin of forewing yellowish. Before the base of the cleft two small spots, one well above the other and separated. In the fringes of the first lobe a black dot at the anal angle, along the termen of the second lobe two small black fringe dots.

MALE GENITALIA.– Valvae moderately wide. Anellus arms two thirds as long as tegumen. Uncus slender, extending beyond margin of tegumen by just over half of its length.

FEMALE GENITALIA.– Ostium slightly excavated. Antrum five times as long as wide. The sclerotised antrum part is as long as the membraneous part of the ductus bursae. In the ductus bursae a sclerotised plate.

DISTRIBUTION.– Throughout the area, extending into Asia Minor. The species has been recorded from the Nearctic region.

BIOLOGY.– The moth flies in June and July, and again in August. The hostplant is *Veronica chamaedrys* L. (Schmid, 1863; Hannemann, 1977a; Hannemann, 1977b; Emmet, 1979; Buszko, 1986; Arenberger & Jaksic, 1991), which grows on wet meadows, edges of woodlands, marshes and shady slopes. The eggs are deposited on the underside of a leaf. The larva feeds on the leaves, flowers and seeds, and overwinters in an excavated part of the stem.

REMARKS.– LHomme (1939) mentions *Convolvulus arvensis* L. and *Gratiola officinalis* L. as hostplants. The former may be a hostplant, but the present species is often confused with reddish forms of *Emmelina monodactyla* L., whereas the latter record almost certainly refers to the recently recognized *S. gratiolae*.

43. *Stenoptilia mannii* (Zeller, 1852)

Pterophorus mannii Zeller, 1852: 375. Pl. 5: 7. Figs 58, 195.
Stenoptilia megalochra Meyrick, 1927: 570.

DIAGNOSIS.– Wingspan 18-30 mm. Colour reddish-ochreous, with slight darkening at the costa. No markings, but for a small faint discal dot. Terminal fringes of both forewing lobes without black dots.

MALE GENITALIA.– Valvae rather slender. Anellus arms long, four fifths as long as the tegumen. Uncus extending for half its length beyond the tegumen margin.

FEMALE GENITALIA.– Ostium excavated. Antrum five to six times as long as wide. The membraneous part of the ductus bursae as long as the antrum and with a very narrow sclerotised plate.

DISTRIBUTION.– The Balkan countries, extending into Asia Minor and Iraq.

BIOLOGY.– The moth flies in July and August. The hostplant is unknown.

REMARKS.– Records from Morocco need confirmation, as these specimens may belong to *S. friedeli* Arenberger, 1984.

44. *Stenoptilia veronicae* Karvonen, 1932

Stenoptilia veronicae Karvonen, 1932: 79. Pl. 5: 8. Figs 59, 196.

DIAGNOSIS.– Wingspan 23-26 mm. Colour cinnamon-brown, speckled white, between the disc and the dorsum yellow-brown. Along the costa dark scales which form a line; at the base of the cleft a distinct double spot (larger than in *S. bipunctidactyla*), in general confluent, and extending towards the wing base as two short ill-defined black lines; an ill-defined small longitudinal spot in the centre of the first lobe. Terminal fringes basally white; the first lobe with a small black dot at the anal angle; the second lobe with three small black dots.

MALE GENITALIA.– Valvae more compressed than in *S. bipunctidactyla*. Anellus arms two thirds as long as the tegumen. Uncus slender, extending for half its length beyond margin of tegumen. Aedeagus shorter than in *S. bipunctidactyla*.

FEMALE GENITALIA.– Ostium excavated. The antrum three times as long as wide. The sclerotised part of the antrum as long as the membraneous part of the ductus bursae, the latter containing a long sclerotised plate.

DISTRIBUTION.– Norway, Sweden, Finland, Baltic states, Poland.

BIOLOGY.– The moth flies from the beginning of June to mid-July. The hostplant is *Veronica longifolia* L. (Karvonen, 1932; Buszko, 1986), which grows on periodically flooded brook-side meadows and peat bogs along rivers.

45. *Stenoptilia bipunctidactyla* (Scopoli, 1763)

Phalaena bipunctidactyla Scopoli, 1763: 257.
Alucita mictodactyla [Denis & Schiffermüller], 1775: 320.
Pterophorus hodgkinsonii Gregson, 1868: 178.
Pterophorus hirundodactylus Gregson, 1871: 364.

Pl. 5: 9. Figs 60, 197.

Taxa considered to belong to the present species but the current status of which is uncertain and needs verification:

Pterophorus plagiodactylus Stainton, 1851: 28.
Pterophorus serotinus Zeller, 1852: 361.
Pterophorus scabiodactylus Gregson, 1871: 363.
Stenoptilia succisae Gibeaux & Nel, 1991: 104.

DIAGNOSIS.– Wingspan 17-25 mm. Colour dark brown. Markings black-brown. Forewing with a discal spot; a double spot at the base of the cleft, usually fused; a longitudinal spot in the centre of the first lobe. In the fringes of the first lobe a black spot at the anal angle, and at the termen of the second lobe two spots, one at the apex and one at two thirds, surrounded by numerous, less well developed, black-grey scales which forms an almost continuous line.

MALE GENITALIA.– Valvae rather slender. Anellus arms two thirds as long as tegumen, variably wide at the apex. Uncus small, just reaching beyond the tegumen margin.

FEMALE GENITALIA.– Ostium slightly excavated. Antrum three times as long as wide. Membraneous part of ductus bursae one and a half times as long as the antrum, containing a sclerotised plate.

DISTRIBUTION.– Western, Central, South and South-East Europe, to the north as far as to central Scandinavia. The area extends into Asia Minor, Syria, (possibly Iran) and Northern Africa.

BIOLOGY.– The moth flies, depending on latitude and altitude, from April to October, in successive generations. The hostplants are *Knautia arvensis* L. (Emmet, 1979; Buszko, 1986; Nel, 1987b; Arenberger & Jaksic, 1991; Gibeaux & Nel, 1991), *Scabiosa columbaria* L. (Hannemann, 1977b; Buszko, 1986; Nel, 1989a; Bigot et al., 1990; Gibeaux & Nel, 1991), *S. pyrenaica* Allioni (Gibeaux & Nel, 1991), *Linaria vulgaris* Miller (Haggett, 1956), *Succisa pratensis* Moench. (Doets, 1946; Emmet, 1979; Gibeaux & Nel, 1991) and *Misopates* (=*Antirrhinum*) *orontium* L. The larva feeds on the flowers and seeds. Pupation on a stem.

REMARKS.– According to literature, a great number of possible hostplants may be added, including: *Scabiosa arvensis* L. (Mitterberger, 1912), *S. lucida* Villiers (Nel, 1987b), *Galium mollugo* L. (South, 1881; Mitterberger, 1912), *Saxifraga granulata* L. , *Euphrasia* sp., *L. cymbalaria* Mill. (Mitterberger, 1912; Nel, 1987b), *Scutellaria galericulata* L. (Mitterberger, 1912), *Gentiana asclepiadea* L., *Coris monspeliensis* L. and *Globularia alypum* L. (Nel, 1987b). These hostplants need verification. For the status of the uncertain species see remarks under *S. aridus*.

46. *Stenoptilia aridus* (Zeller, 1847)

Pterophorus aridus Zeller, 1847: 904.   Pl. 5: 10. Fig. 198.
Stenoptilia stigmatodactyla var. *grisescens* Schawerda, 1933: 74.
Stenoptilia csanadyi Gozmány, 1959: 367.
Stenoptilia gallobritannidactyla Gibeaux, 1985a: 248.

Taxa considered to belong to the present species but the current status of which is uncertain and needs verification:

Stenoptilia mimula Gibeaux, 1985a: 246.
Stenoptilia picardi Gibeaux, 1986: 328.

DIAGNOSIS.– Wingspan 13-18 mm. Colour ochreous brown. The pattern dark brown, basic elements as in *S. bipunctidactyla*.

MALE GENITALIA.– As in *S. bipunctidactyla*.

FEMALE GENITALIA.– Ostium asymmetrical. Antrum three and a half times as long as wide, gradually narrowing.

DISTRIBUTION.– Mediterranean area and northern Africa.

BIOLOGY.– The moth flies from April to October. Recorded hostplants are *Succisa pratensis* Moench (Nel, 1987b), *Erinus alpinus* L., *Linaria origanifolia* DC. (Nel, 1987b), *Antirrhinum orontium* L. (Nel, 1987b) and *Coris monspeliensis* L. (Nel, 1987b; Bigot et al., 1990).

REMARKS.– The distribution and hostplant records of the present species are given tentatively because of confusion with the preceding species and the uncertain status of the other taxa mentioned with both *S. bipunctidactyla* and *S. aridus*. The distribution scheme is not given, as the records are mixed with those of *S. bipunctidactyla*. There is great need for a specific research, including SEM analysis and pheromone typing, to establish all species in this difficult complex.

47. *Stenoptilia elkefi* Arenberger, 1984

Stenoptilia elkefi Arenberger, 1984: 10.   Pl. 6: 1. Figs 61, 199.

DIAGNOSIS.– Wingspan 17-19 mm. Colour ochreous brown, mixed with numerous white scales. At the base of the cleft a double spot, the top spot only indicated by a few sparse scales. Fringes pale brown, small black spot in the fringes at the anal angle of the first lobe, and at the apex and centre of the second lobe.

MALE GENITALIA.– Valvae of moderate size, sacculus spindle-shaped, cucullus curved around middle. Anellus arms half as long as tegumen. Top of tegumen flattened. Uncus just beyond the margin of the tegumen.

FEMALE GENITALIA.– Ostium excavated. Antrum saucer-like, wide, with a weakly sclerotised spade-like distal extension. Ductus bursae long, five times as long as the antrum, with a long sclerotised plate in middle.

DISTRIBUTION.– Mediterranean area, Asia Minor, Jordan.

BIOLOGY.– The moth flies from May to September, probably in three generations. The hostplant is *Scabiosa atropurpurea* L. (Nel, 1987b; Arenberger & Jaksic, 1991). The larva feeds on the flowers, flower-buds and seeds, like *S. bipunctidactyla.*

48. *Stenoptilia lucasi* Arenberger, 1990

Stenoptilia lucasi Arenberger, 1990c: 101.   Pl. 6: 2. Figs 62, 200.

DIAGNOSIS.– Wingspan 16-20 mm. Colour brown, speckled with white scales. Markings black-brown: a double spot at the base of the cleft; a discal spot; a longitudinal spot in the centre of the first lobe, and obliquely above this a further one near the costa. In the fringes the small black spot at the anal angle of the first lobe hardly developed, at the termen of the second lobe a small apical spot and one at two thirds.

MALE GENITALIA.– Valvae moderately developed, the apical process rather short. Anellus arms long, five sixths as long as tegumen. Uncus well developed, extending for half its length beyond the margin of the tegumen.

FEMALE GENITALIA.– Ostium smoothly excavated. Antrum gradually narrowing, three times as long as wide. Membraneous part of the ductus bursae two and a half times as long as antrum, with a large sclerotised plate in middle.

DISTRIBUTION.– Greece, Asia Minor.

BIOLOGY.– The moth flies in July. The hostplant is unknown.

49. *Stenoptilia annadactyla* Sutter, 1988

Stenoptilia annadactyla Sutter, 1988: 181.   Pl. 6: 3. Figs 63, 201.
Stenoptilia annickana Gibeaux, 1989a: 222.

DIAGNOSIS.– Wingspan 17-24 mm. Colour brown-grey, at the dorsum pale brown to yellowish. The pair of spots at the base of the cleft often confluent, one above the other. Basally of the costal spot at the cleft often a small longitudinal spot towards the base of the wing; a discal spot; a longitudinal spot in the centre of the first lobe. In the fringes of the first lobe a black spot at the anal angle.

MALE GENITALIA.– Valvae rather stout. Sacculus spindle-shaped, basally wider than distally. Anellus arms moderately wide (more slender than in *S. bipunctidactyla*), two thirds of tegumen length. Uncus small, extending beyond the margin of the slightly excavated tegumen by half of its length. Aedeagus rather short and stout.

FEMALE GENITALIA.– Ostium slightly excavated. Antrum with almost parallel margins, one and a half times as long as wide. Membraneous part of ductus bursae twice as long as antrum, with a sclerotised plate in half of its length.

DISTRIBUTION.– Western and Central Europe.

BIOLOGY.– The moth flies from the beginning of June to mid-September. Gibeaux (1989a) mentions a bred specimen from mid-October. The hostplant is *Scabiosa columbaria* L. (Gibeaux, 1989a; Gibeaux & Nel, 1991). The larva feeds on the flower-heads.

50. *Stenoptilia pelidnodactyla* (Stein, 1837)

Alucita pelidnodactyla Stein, 1837: 98.   Pl. 6: 4. Figs 64, 202.

Taxa which may be considered to belong to the present species the status of which is uncertain and needs verification:

Stenoptilia pelidnodactyla ssp. *alpinalis* Burmann, 1954: 187.
Stenoptilia bigoti Gibeaux, 1986: 332.
Stenoptilia gibeauxi Nel, 1989c: 468.
Stenoptilia cerdanica Nel & Gibeaux, 1990: 134.
Stenoptilia cebennica Nel & Gibeaux, 1990: 134.
Stenoptilia mercantourica Nel & Gibeaux, 1990: 134.

DIAGNOSIS.– Wingspan 16-25 mm. Colour grey. Markings black: a double spot at the base of the cleft, one above the other, the dorsal one larger than the costal one; a discal spot; a longitudinal spot in the centre of the first lobe and dark scales along the costa, almost forming a costal line. The fringes contain a black basal spot at the anal angle of the first and second lobes, and spots also at the apex of the second lobe and at the termen at two thirds of the second lobe.

MALE GENITALIA.– Valvae with almost parallel margins, the tip well developed. Sacculus distinctly wider basally than distally. Anellus arms stout, two thirds as long as tegumen length. Uncus just extending beyond the slightly excavated distal tegumen margin. Aedeagus stout, of moderate length.

FEMALE GENITALIA.– Ostium excavated. Antrum one and a half times as long as wide. Membraneous part of ductus bursae twice as long as the antrum.

DISTRIBUTION.– In mountainous areas in Central and southern Europe; in Scandinavia at lower altitudes.

BIOLOGY.– The moth flies, depending on the latitude and altitude, from April to July and again in September. The hostplants are *Saxifraga granulata* L. (Gozmány, 1962; Hannemann, 1977b; Buszko, 1986; Nel, 1986d; Bigot et al., 1990), *S. bryoides* L. (Hannemann, 1977b), *S. pedemontana* All. ssp. *cervicornis* (Viv.) (Nel, 1989c; Nel & Gibeaux, 1990; Nel, 1991), *S. moschata* Wulfen (Burmann, 1954; Hannemann, 1977b), *S. nervosa* Lapeyr. (Nel & Gibeaux, 1990), *S. prostii* Sternb. (Nel & Gibeaux, 1990), *S. exarata* Vill. (Nel & Gibeaux, 1990), *S. aquatica* Lapeyr. (Nel & Gibeaux, 1990), *S. geranioides* L. (Nel & Gibeaux, 1990) and *Plantago sempervirens* Crantz. (Gibeaux, 1986; Nel, 1987b; Bigot et al., 1990). The first of these grows on open spots in woodlands and on grassy hills, the last two occur in alpine regions. Other recorded, but unverified, hostplants are *Globularia* sp., *Gentiana verna* L. and *G. acaulis* L. (= *G. kochiana* Perr. and Song.). The larva feeds on the flower-buds, the flowers and seeds, occasionally the leaves. Pupation usually on the stems of the hostplant, but the ssp. *alpinalis* tends to pupate under surrounding stones as well. The pupal stage lasts approximately 14 days.

REMARKS.- The species described by Nel and Nel & Gibeaux have external characters resembling the present species, particularly with regard to the fringe spots on the forewings. The genital differences are minute and are, in my opinion, within the range of variation of the present species. The breeding experiments of these species are, however, so interesting that separate status may be considered valid after further research. See remarks under *S. aridus*.

50a. *Stenoptilia brigantiensis* Nel & Gibeaux, 1992

Stenoptilia brigantiensis Nel & Gibeaux, 1992: 53.
Stenoptilia buvati Nel & Gibeaux, 1992: 56.

These two 'species' are neither described, nor illustrated in this review. They were only recently recognized as belonging to a distinct group.

DISTRIBUTION.- French Alps.

BIOLOGY.- The moth flies in July. The hostplants are *Saxifraga exarata* Vill. and *S. moschata* Wulf. (Nel & Gibeaux, 1992).

51. *Stenoptilia reisseri* Rebel, 1935

Stenoptilia reisseri Rebel, 1935: 11.   Pl. 6: 5. Figs 65, 203.

DIAGNOSIS.- Wingspan 18-20 mm. Colour chalk-white. Markings grey-black: a large double spot at the base of the cleft; a discal spot; a darkening along the costa, up to the base of the cleft, and a longitudinal spot in the centre of the first lobe. A black spot in the fringes of the first lobe at the anal angle, and other three spots on the second lobe at the apex and at two thirds of termen.

MALE GENITALIA.- Valvae short and stout. Sacculus spindle-shaped, the protrusion before the apex rounded. Anellus arms short, three fifths as long as tegumen. Uncus slender, and projects well beyond the margin of the tegumen. Aedeagus of moderate size.

FEMALE GENITALIA.- Ostium excavated, with rounded lateral margins. Antrum with parallel margins, three times as long as wide. Membraneous part of ductus bursae three times as long as antrum, with a short, broad sclerotised plate.

57

DISTRIBUTION.– Mountains of West-Central Spain.

BIOLOGY.– The moth flies in July and the beginning of August, at altitudes of 1500-2200 m. The hostplant is unknown.

REMARKS.– The record of the species in the Sierra Nevada (Gielis, 1988) is incorrect, and refers to the following species.

52. *Stenoptilia hahni* Arenberger, 1989

Stenoptilia hahni Arenberger, 1989: 327.   Pl. 6: 6. Figs 66, 204.

DIAGNOSIS.– Wingspan 19-20 mm. Colour coffee-brown, markings dark brown. A costal line, which widens towards, and reaches the base of the cleft; a large spot at the base of the cleft; a smaller discal spot; a longitudinal spot centrally in the first lobe reaches both costa and dorsum and is margined by a white, oblique line, continuing onto the second lobe. In the fringes, black dots are present at the apex and the anal angle of both lobes and centrally at the termen of the second lobe.

MALE GENITALIA.– Valvae more elongated than in *S. reisseri*, with a more regular spindle-shape of the sacculus. The protrusion near the apex of the valvae elongated, flattened. Anellus arms stout and short, half as long as tegumen. Uncus stout and wide, two thirds of its length is beyond the tegumen margin. Aedeagus stout, rather short.

FEMALE GENITALIA.– Ostium rounded. Antrum funnel-shaped, with a narrow proximal end, as long as wide. Membraneous part of ductus bursae three to four times as long as the antrum, with a long and narrow sclerotised plate.

DISTRIBUTION.– Mountains of southern Spain.

BIOLOGY.– The moth flies in July and the beginning of August, at altitudes of 1600-2000 m. The hostplant is unknown.

REMARKS.– See *S. reisseri*.

53. *Stenoptilia millieridactyla* (Bruand, 1861)

Pterophorus millieridactyla Bruand, 1861: 36.   Pl. 6: 7. Figs 67, 205.
Stenoptilia saxifragae Fletcher & Pierce, 1940: 25.

DIAGNOSIS.– Wingspan 17-20 mm. Colour grey-brown, between the disc and the dorsum pale reddish-brown. Markings black: a double spot at the base of the cleft, usually hardly confluent; a small discal spot; a longitudinal spot in the dorsal half of the first lobe; also longitudinally grouped scales forming two indistinct rows in the centre of the second lobe. Two small black spots in the fringes of the termen of the first lobe and a spot at the anal angle; in the fringes of the second lobe a basal row of black scales, which are interrupted at middle and at three quarters of termen.

MALE GENITALIA.– Valvae with almost parallel margins, rather slender. Anellus arms slender, three fourths as long as tegumen length, with club-like apices (differentiating the species from all related ones). Uncus stout, extending for half its length beyond the margin of the tegumen.

FEMALE GENITALIA.– Ostium rounded. Antrum funnel-shaped, with curved margins (and not straight as in *S. hahni*). Membraneous part of ductus bursae five to six times as long as antrum, with a very long, slender sclerotised plate.

DISTRIBUTION.– West and South-West Europe.

BIOLOGY.– The moth flies from May to September, probably in two generations. The hostplants are *Saxifraga continentalis* Engler & Irmescher (Nel, 1986d; Nel & Gibeaux, 1990), *S. granulata* L. (Nel, 1986d) and *S. hypnoides* L. (Emmet, 1979). The larva feeds in the petioles of the leaves, later on the leaves. Pupation on the hostplant.

54. *Stenoptilia islandicus* (Staudinger, 1857)

Pterophorus islandicus Staudinger, 1857: 280.   Pl. 6: 8. Figs 68, 206.
Pterophorus pelidnodactylus var. *borealis* Wocke, 1864: 217.

DIAGNOSIS.– Wingspan 17-19 mm. Colour brownish grey, with a purplish gloss. Markings dark grey: a large, confluent double spot at the base of the cleft; a small discal spot; a longitudinal spot in the dorsal half of the first lobe, and some dark subterminal scales in the second lobe. The fringes of the first lobe with a black spot at the anal angle; at the termen of the second lobe an anal, apical and mid-terminal spot.

MALE GENITALIA.– Valvae rounded, rather stout. Anellus arms stout, wide, two thirds as long as tegumen. Uncus short, by half its length extending beyond the margin of the tegumen.

FEMALE GENITALIA.– Ostium wavy excavated, with a small lateral extension. Antrum gradually narrowing, twice as long as wide. The membraneous part of the ductus bursae twice as long as the antrum, with a large sclerotised plate.

DISTRIBUTION.– Iceland, Scotland, Norway, Sweden.

BIOLOGY.– The moth flies in June and July. The hostplants are *Saxifraga caespitosa* L. and *S. adscendens* L. (Aarvik et al., 1986). The larva feeds at night, by day hiding under the leaves of the hostplant.

REMARKS.– The status of the f. *borealis* Wocke is under discussion. Wolff (1964) considers it a synonym of the present species; Zagulajev (1986) treats it as a bona species and so do Nel & Gibeaux (1990). In the absence of the type specimen of this form, the present status is probably best retained.

55. *Stenoptilia parnasia* Arenberger, 1986

Stenoptilia parnasia Arenberger, 1986: 80.   Pl. 6: 9. Figs 69, 207.

DIAGNOSIS.– Wingspan 16-17 mm. Colour coffee-brown. Along the costa dark brown as far as just beyond the base of the cleft, here interrupted with white; a double, confluent spot at the base of the cleft, reaching up to the costa; in the centre of the first lobe a longitudinal spot; some sparse dark scales in the centre of the second lobe. Fringes with a small black dot at the anal angle of the first lobe and an anal, apical and mid-terminal dot on the second lobe.

MALE GENITALIA.– Valvae spindle-shaped. The process before the apex has a characteristic protrusion, extending parallel to the apex. Anellus arms short, stout, and half as
long as the tegumen. Uncus small, only slightly projecting beyond the margin of the
tegumen.

FEMALE GENITALIA.– Ostium excavated (not as deeply as in *S. elkefi*). Antrum saucer-like,
wide. Membraneous part of ductus bursae long, ten times as long as the antrum, with
a long sclerotised plate in middle.

DISTRIBUTION.– The mainland of Greece.

BIOLOGY.– The moth flies in July and August. The hostplant is unknown.

56. *Stenoptilia coprodactylus* (Stainton, 1851)

Pterophorus coprodactylus Stainton, 1851: 28.   Pl. 6: 10. Figs 70, 208.
Stenoptilia zalocrossa Meyrick, 1907: 146.

The following recently described species, the status of which needs confirmation, is tentatively
arranged here.

Stenoptilia pseudocoprodactyla Gibeaux, 1992: 467.

DIAGNOSIS.– Wingspan 18-27 mm. Colour ochreous-brown. Markings dark brown: a separated double spot well before the base of the cleft; a small discal spot; a longitudinal
spot in the centre of the first lobe; some dark scales subterminally in the second lobe.
In the fringes of the anal angle of the first lobe a small black spot, and small spots at
the anal angle, the apex and mid-termen of the second lobe.

MALE GENITALIA.– Valvae rather slender, with an extended apex. Anellus arms two thirds
as long as the tegumen. Tegumen distally with a membraneous extension, and slightly
excavated in middle. Uncus small, slender, within the margin of the tegumen.

FEMALE GENITALIA.– Ostium excavated. Antrum gradually narrowing, one and a half times
as long as wide. The membraneous part of the ductus bursae one and a half times as
long as antrum, with a stout sclerotised plate in middle.

DISTRIBUTION.– Mountainous regions of South, Central and South-East Europe.

BIOLOGY.– The moth flies in May and June, later in July and August. The hostplants are
Gentiana verna L. (Mitterberger, 1912; Gozmány, 1962; Hannemann, 1977b; Buszko, 1986; Bigot et al., 1990), *G. acaulis* L. (= *G. kochiana* Perr. & Song.) (Nel, 1984)
and *G. lutea* L. (Hannemann, 1977b; Arenberger & Jaksic, 1991). The larva feeds on
the flower-buds, flowers and seed-heads. Pupation usually on the stem.

REMARKS.– The recorded hostplant *G. lutea* may be incorrect as a result of confusing the
species with *S. lutescens* (H.-S.).

57. *Stenoptilia lutescens* (Herrich-Schäffer, 1855)

Pterophorus lutescens Herrich-Schäffer, 1855: 377.   Pl. 6: 11. Figs 71, 209.
Mimaesoptilus arvernicus Peyerimhoff, 1875: 515.
Stenoptilia grandis Chapman, 1908: 318.

DIAGNOSIS.– Wingspan 25-32 mm. Colour reddish grey-brown. Markings consist of a double, often weakly confluent, spot at the base of the cleft, the costal spot slightly displaced towards the wing base; a discal spot; irregularly distributed dark scales in the second lobe and a longitudinal spot in the centre of the first lobe. This spot in the first lobe is characteristically margined terminally by an oblique white line which forms an acute angle with the termen. Fringes with small black dots at the anal angles of both lobes, and at the second lobe at the termen at two thirds and at the apex.

MALE GENITALIA.– Valvae basally wide, distally narrower. The saccular process just before the apex is rather small. Anellus arms slender-tipped, half as long as the tegumen. Uncus stout and short, the apex reaches the margin of the tegumen. Aedeagus rather long and slender.

FEMALE GENITALIA.– Ostium almost flat. Antrum one and a half times as long as wide. The membraneous part of the ductus bursae four times as long as the antrum, with a slender sclerotised plate for over half its length.

DISTRIBUTION.– Mountainous areas in Central and South-West Europe.

BIOLOGY.– The moth flies from July to mid-August, at altitudes of 1300-2100 m. The hostplant is *Gentiana lutea* L. (Chapman, 1908; Bigot et al., 1990). The feeding habits have not yet been described. Pupae are found near the mid-rib on the upper-side of leaves.

58. *Stenoptilia nepetellae* Bigot & Picard, 1983

Stenoptilia nepetellae Bigot & Picard, 1983: 21.   Pl. 7: 1; pl. 15: 5, 6. Figs 72, 210.
Stenoptilia cyrnea Nel, 1991: 170.

DIAGNOSIS.– Wingspan 25-29 mm. Colour ochreous grey-brown. Markings black-grey: a double spot just before the base of the cleft, the costal spot markedly smaller than the dorsal spot; a small discal spot; some dark scales in the centre of the first lobe. In the fringes a small black spot at the anal angle of the first lobe, of the second lobe a small apical spot and one at two thirds of the termen. Between this last spot and the anal angle of the second lobe, a faint darkening of the fringes, but no actual spot.

MALE GENITALIA.– Valvae with some widening near the end of the sacculus. The saccular process just before the apex of the valve is large and rounded. Anellus arms slender, rather long, three fourths as long as the tegumen. Uncus slender and extending for nearly two thirds of its length beyond the margin of the tegumen. Aedeagus long and slender.

FEMALE GENITALIA.– Ostium excavated. Antrum gradually narrowing, four times as long as wide. Membraneous part of the ductus bursae about as long as antrum, with a small sclerotised plate.

DISTRIBUTION.– Western Alps, the Pyrenees and Corsica.

BIOLOGY.– The moth flies in July and August, at altitudes of 1200-2000 m. The host-plants are *Nepeta nepetella* L. and *N. agrestis* Loisel. The larva feeds on the young shoots and leaves. If flower-buds and flowers are developed, they are eaten as well. Pupation on a leaf or stem. The pupal stage lasts 15-21 days.

REMARKS.– The recently described species from Corsica (Nel, 1991) has the wing markings and genital structures as *nepetellae*. Further see remarks *Crombrugghia distans*.

59. *Stenoptilia stigmatodactylus* (Zeller, 1852)

Pterophorus stigmatodactylus Zeller, 1852: 374.   Pl. 7: 2. Figs 73, 211.
Pterophorus oreodactylus Zeller, 1852: 374.

DIAGNOSIS.– Wingspan 17-23 mm. Colour ochreous brown to ochreous grey-brown, between the disc and dorsum more ochreous tinged. Markings dark brown: a double spot before the base of the cleft (the costal spot smaller and basally displaced); a discal spot. In the fringes of the first lobe an anal spot, and at the second lobe an anal, apical and mid-terminal spot.

MALE GENITALIA.– Valvae of moderate size with almost parallel margins. The saccular process less extended than in the previous species. Anellus arms very long, five sixths as long as the tegumen. Uncus small and just reaches beyond the margin of the tegumen. Aedeagus very long.

FEMALE GENITALIA.– Ostium excavated. Antrum gradually narrowing, five to six times as long as wide. Membraneous part of ductus bursae as long as antrum, with a long and slender sclerotised plate.

DISTRIBUTION.– Western, Central and Mediterranean Europe, extending into Asia Minor and North Africa.

BIOLOGY.– The moth flies from June to September, probably in two generations. The hostplants are *Thymus vulgaris* L. (Nel, 1987b) and *Scabiosa ochroleuca* L. (Gozmány, 1962; Hannemann, 1977b). The larva feeds on the flower-buds, flowers and unripe seeds. Pupation along the stem; the pupa is covered by a loose spinning. Buszko (1986) also mentions *Scabiosa lucida* Vill. and *Knautia arvensis* (L.) as hostplants.

60. *Stenoptilia stigmatoides* Sutter & Skyva, 1992

Stenoptilia stigmatoides Sutter & Skyva, 1992: 81.   Pl. 7: 3. Figs 74, 212.

DIAGNOSIS.– External characters as in *S. stigmatodactylus*.

Male genitalia.– Valvae and anellus arms as in *S. stigmatodactylus*. Aedeagus shorter, more stout.

FEMALE GENITALIA.– Ostium rather flat. Antrum with almost parallel margins, two and a half times as long as wide.

DISTRIBUTION.– Slovakia.

BIOLOGY.– The moth flies from the end of May to mid-September.

REMARKS.– Easily confused with *S. stigmatodactylus*, but clearly differing in the female genitalia.

61. *Stenoptilia zophodactylus* (Duponchel, 1840)

Pterophorus zophodactylus Duponchel [in: Godart], 1840b: 668. Pl. 7: 4; pl. 15: 9-11.
Pterophorus loewii Zeller, 1847: 38. Figs 13, 75, 213.
Pterophorus canalis Walker, 1864: 944.
Mimeseoptilus semicostata Zeller, 1873: 323.

DIAGNOSIS.– Wingspan 16-23 mm. Colour pale brown. Markings of dark-brown scales: a distal spot and two spots before the base of the cleft: the costal one a little more basally placed than the dorsal one. Sparse dark scales along the costa and dorsum, and in the first lobe. In fringes at termen of the first lobe two scale groups, and three such groups at the second lobe.

VARIATION.– The intensity of the markings is variable. The colour may be dark-brown to pale-brown.

MALE GENITALIA.– Valvae with a small saccular process near the apex. Anellus arms half as long as tegumen. Tegumen with two membraneous projections at apex. Uncus small, the apex not exceeding the distal margin of the tegumen. Aedeagus strongly curved.

FEMALE GENITALIA.– Ostium rounded with an acute distal part. Antrum gradually narrowing, three times as long as wide. Membraneous part of ductus bursae one and a half times as long as antrum, with a large sclerotised plate for nearly all of its length.

DISTRIBUTION.– Western, Central, and South-East Europe, the Mediterranean area, extending into Asia Minor and North Africa. The moth is known from South Africa, India, and Australia, as well as from the Neotropical and Nearctic regions.

BIOLOGY.– The moth flies in Europe from April to November. The number of generations varies with the latitude from one in the north to three or four in the southern part of the region. The hostplants are *Centaurium erythraea* Rafn. (=*C. minus* Mönch) (Schmid, 1863; Goury, 1912; Gozmány, 1962; Hannemann, 1977b; Gielis, bred), *C. umbellatum* Gilib. (Buszko, 1986), *C. littoralis* (D. Turner) Gilmour (Goury, 1912; Gielis, bred), *Blackstonia perfoliata* (L.) Hudson (Beirne, 1954; Emmet, 1979) and *Gentianella germanica* Willd. (Gozmány, 1962; Hannemann, 1977b; Arenberger & Jaksic, 1991) in Europe. A North American specimen has been labelled: *Erythraea venusta* A. Gray (Barnes & Lindsey, 1921). The larva feeds on the flowers and seeds of these plants. Pupation along the stem.

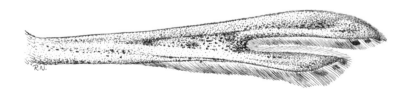

Fig. 13. Forewing of *Stenoptilia zophodactylus*. (R. Nielsen del.).

Buszkoiana Koçak, 1981

Richardia Buszko, 1978: 77; homonym of *Richardia* Robineau-Desvoidy, 1830 (Insecta, Diptera).
 Type species: *Pterophorus capnodactylus* Zeller, 1841; original designation.
Buszkoiana Koçak, 1981: 10; replacement name for *Richardia* Buszko, 1978.

DIAGNOSIS.– Scale-tooth on third lobe of hindwing at apex. Costal triangle ill-defined. Forewing vein R1 present.

MALE GENITALIA.– Valvae symmetrical, lanceolate, apex rounded, sacculus not interrupted, without valvular or saccular processes. Saccus in shape of a sclerotised pentangular plate. Tegumen bilobed. Uncus well developed.

FEMALE GENITALIA.– Antrum sclerotised, centrally ending on the distal margin of the 7th sternite. Ductus bursae without a sclerotised element. The lamina postvaginalis not developed. Signum in bursa copulatrix simple, consisting of a weakly sclerotised section of bursal wall.

DISTRIBUTION.– South-East, Central and West Europe, to the North into southern Scandinavia.

BIOLOGY.– Asteraceae (*Petasites hybridus* G., M. & S.) (Wolff, 1953; Nielsen, 1962).

62. *Buszkoiana capnodactylus* (Zeller, 1841)

Pterophorus capnodactylus Zeller, 1841: 774.   Pl. 7: 5,6. Figs 76, 214.

DIAGNOSIS.– Wingspan male 18-24 mm; female 22-27 mm. Colour of the forewing of the male dark brown to black-brown, the female much paler, ochreous brown, and with a small ochreous spot at the costa near the base of the cleft. Hindwing black-brown.

MALE GENITALIA.– See genus description.

FEMALE GENITALIA.– See genus description.

DISTRIBUTION.– See genus description.

BIOLOGY.– The moth flies in June and July. The hostplant, *Petasites hybridus* G., M. & S. (Nielsen, 1962; Hannemann, 1977a; Biesenbaum, 1987) grows on wet, often shady, places in woodlands, along brooks and in marshes. The larva feeds in the lower parts of the stem, and drops the faecal remains in a spinning through a hole in the stem.

REMARKS.– The species was described from Hungary. Isolated records exist from areas of ex-Yugoslavia near the type locality. In the middle of this century records indicate an expansion of the range in a western, northern and north-western direction.

Cnaemidophorus Wallengren, 1862

Cnaemidophorus Wallengren, 1862: 10.
 Type species: *Alucita rhododactyla* [Denis & Schiffermüller], 1775; monotypy.
Cnemidophorus Zeller, 1867; emendation & homonym.

Eucnemidophorus Wallengren, 1881; unnecessary replacement name for unjustified emendation.

Euenemidophorus Pierce & Metcalfe, 1938; incorrect spelling.

DIAGNOSIS.– Scale-tooth on third lobe of hindwing at apex. Costal triangle well developed. Forewing vein R1 present.

MALE GENITALIA.– Valvae symmetrical, lanceolate, their apex rounded; a large spine points toward the vinculum from the middle of the saccular section (which is not separate from the valve). Saccus minute. Tegumen bilobed. Uncus stout.

FEMALE GENITALIA.– The antrum has a conical shape and ends distad to the distal margin of the 7th sternite. Ductus bursae without a sclerotised segment. The lamina postvaginalis not developed. The signum consists of a double sclerotised plate, covered with minute spiculae.

DISTRIBUTION.– Holarctic.

BIOLOGY.– The hostplant is *Rosa* sp. (Hofmann, 1896).

63. *Cnaemidophorus rhododactyla* ([Denis & Schiffermüller], 1775)

Alucita rhododactyla [Denis & Schiffermüller], 1775: 146.   Pl. 7: 7. Figs 77, 215.
Platyptilia koreana Matsumura, 1931: 1055.

DIAGNOSIS.– Wingspan 18-26 mm. Palpi short and slender with basal segment strongly scaled. Antennae ringed in white and brown. Colour orange-brown. Markings white: at the dorsum at 2/3, a discal spot; an angulated transverse fascia at the base of the cleft, angulated toward the base of the wing. This last fascia basally margined by black scales.

MALE GENITALIA.– See genus description.

FEMALE GENITALIA.– See genus description.

DISTRIBUTION.– Holarctic, except for the far north of the region.

BIOLOGY.– The moth flies from June to August. The hostplant is *Rosa canina* L. (Porritt, 1875; Gozmány, 1962; Emmet, 1979), *R. spinosissima* L. (Doets, 1950) and other *Rosa* spp. (Hofmann, 1896), among them also cultivated species. Yano (1963) records *Rosa rugosa* Thunb. from Japan. The larva feeds in a loose spin in the terminal shoot. First the developing leaves are attacked, later the flower buds and the flowers are eaten. Pupation on the plant.

REMARKS.– The distribution of the species does not follow the distribution of its hostplant. There seems to be a relation to the calcareous content of the soil, as noticed for *Capperia trichodactyla* by Adamczewski (1951).

Marasmarcha Meyrick, 1886

Marasmarcha Meyrick, 1886: 11.
  Type species: *Alucita phaeodactyla* Hübner, [1813]: pl. 3, figs 14, 15; subsequent designation (Tutt, 1906).

DIAGNOSIS.– Third lobe of the hindwing without a scale-tooth. Forewing markings reduced to some pale transverse fasciae, costal triangle not developed. Forewing vein R1 absent.

MALE GENITALIA.– Valvae lanceolate to trapezoid. The pointed spine arising from mid-sacculus very slender, sometimes split into two prolonged structures (which may occasionally be asymmetrical). Sacculus not lobed. Saccus small. Tegumen bilobed, rather small. Uncus rather broad. (The genitalia tend to be slightly asymmetrical).

FEMALE GENITALIA.– Antrum has a plate-like shape, appears as a prolongation of the distal margin of the 7th sternite, and ends medially at the distal margin of this sternite. Ductus bursae without a sclerotised segment. The lamina postvaginalis may be developed as a sclerotised ridge. Signum obsolete (ill-marked when present).

DISTRIBUTION.– Palaearctic region.

BIOLOGY.– The hostplants are Fabaceae: *Ononis* spp. (Zeller, 1852; Frey, 1856; Schwarz, 1953).

64. *Marasmarcha lunaedactyla* (Haworth, 1811)

Alucita lunaedactyla Haworth, 1811: 477.　　　　　　　　Pl. 7: 8. Figs 78, 216.
Alucita phaeodactyla Hübner, [1813]: t.4, figs 14, 15.
Pterophorus agrorum Herrich-Schäffer, 1855: 378.
Marasmarcha altaica Krulikowskij, 1906: 51.
Marasmarcha agrorum tuttodactyla Chapman, 1906: 178.

DIAGNOSIS.– Wingspan 18-22 mm. Colour dark ferrugineous brown. Markings confined to a pale, nicked, incomplete transverse band at the base of the cleft.

MALE GENITALIA.– Valvae symmetrical. In the basal half of each valve a pair of strongly curved spines which are semicircular in cross-section.

FEMALE GENITALIA.– Ostium in shape of an oval sclerotised plate, with the antrum ending centrally in this plate. Lamina antevaginalis hardly indicated.

DISTRIBUTION.– Europe, northwards into Sweden.

BIOLOGY.– The moth flies from June to August. The larvae feed on *Ononis repens* L. (Mitterberger, 1912; Hannemann, 1977b; Emmet, 1979), *O. natrix* L. (Nel, 1986c), *O. rotundifolia* L. (Nel, 1986c), *O. arvensis* L. (Hannemann, 1977b; Buszko, 1986) and *O. spinosa* L. (Beirne, 1954; Hannemann, 1977b; Emmet, 1979) which grows in dry meadows and along roadsides. The larva feeds on the flowers and shoots and pupates along a shoot or on a leaf.

65. *Marasmarcha fauna* (Millière, 1871)

Mimaeseoptilus fauna Millière, 1871: 380.　　　　　　Pl. 7: 9; pl. 15: 7, 8. Figs 79, 217.

DIAGNOSIS.– Wingspan 16-20 mm. Colour ochreous to orange ferrugineous. The forewing has a complete transverse band at the base of the cleft.

MALE GENITALIA.– Valvae symmetrical. The pair of spines in the basal half of the valvae are symmetrical, but flattened (not semi-circular) in a cross-section.

FEMALE GENITALIA.– Ostium plate rounded with an angulated top (like a cardinals hat). Lamina antevaginalis moderately sclerotised.

DISTRIBUTION.– South-eastern France.

BIOLOGY.– The moth flies in June and July. The hostplant is *Ononis spinosa* L. The author collected larvae and pupae of the species on this plant and adults were on the wing at the same time.

REMARKS.– The synonymy of this species with *M. lunaedactyla*, as proposed by Gibeaux (1990b), is a misinterpretation. The smoothly curved saccular spines are interpreted by him as identical with those of the previous species.

66. *Marasmarcha oxydactylus* (Staudinger, 1859)

Pterophorus oxydactylus Staudinger, 1859: 258.   Pl. 7: 10. Figs 80, 218.
Marasmarcha wullschlegeli Müller-Rutz [in: Vorbrodt & Müller-Rutz], 1914: 446.

DIAGNOSIS.– Wingspan 18-22 mm. Colour brown to ferrugineous ochreous. The forewing with a nicked, complete transverse band at the base of the cleft.

MALE GENITALIA.– Valvae slightly asymmetrical. The left valve with a pair of rather long curved spines; the right valve with a pair of shorter spines.

FEMALE GENITALIA.– Ostium oval with an excavated top. The lamina antevaginalis shaped as a strongly sclerotised ridge.

VARIATION.– In the Mediterranean area extremely pale specimens occur, which may be cream-white.

DISTRIBUTION.– Central and South-West Europe.

BIOLOGY.– The moth flies from May to August. In the mountains of Central Europe an elevation of 2000 m is reached. The species has been recorded bred on *Ononis rotundifolia* L. (Burmann, 1986; Nel, 1986c).

Geina Tutt, 1907

Geina Tutt, 1907: 411.
Type species: *Phalaena Alucita didactyla* Linnaeus, 1758; monotypy.

DIAGNOSIS.– Palpi simple, without a hair-brush along the third segment. The first forewing lobe acute, the second lobe terminally sinuate. On the forewing lobes two transverse white lines. A terminal scale-tooth at the third lobe of the hindwing.

MALE GENITALIA.– Valvae symmetrical, simple, gradually narrowing, without attached processes. Uncus well developed. Aedeagus straight, without cornuti.

FEMALE GENITALIA.– Ostium small. Ductus bursae narrowly tube-like. Bursa copulatrix vesicular, without a signum.

DISTRIBUTION.– The genus is known from the Holarctic region. *G. didactyla* L. occurs in the Palearctic area. In North America another four species are reported.

67. *Geina didactyla* (Linnaeus, 1758)

Phalaena Alucita didactyla Linnaeus, 1758: 542.　　　　　　Pl. 8: 1. Figs 8, 81, 219.
Pterophorus brunneodactylus Millière, 1854: 65.

DIAGNOSIS.– Wingspan 17-23 mm. See genus description. The colour is bright ferrugineous brown. The scale-tooth at the third lobe of the hindwing is well developed and extends on the costa and dorsum in equal parts.

MALE GENITALIA.– See genus characters.

FEMALE GENITALIA.– See genus characters.

DISTRIBUTION.– Europe, except for the far north of the area and the British Isles, extending to the east as far as to Asia Minor and European Russia.

BIOLOGY.– The moth flies from June to the beginning of August. The hostplants are *Geum rivale* L. (Schleich, 1864; Hering, 1891; Hannemann, 1977b; Buszko, 1986) *G. urbanum* L. (Hannemann, 1977b; Buszko, 1986) and *Potentilla rupestris* L. (Adamczewski, 1951; Gozmány, 1962; Hannemann, 1977b; Buszko, 1986; Nel, 1986c). The larva feeds on the flower-buds and flowers. In the buds a lateral hole is bored and the contents eaten. In absence of flowers, leaves are accepted. The colour of the larva is variable, and depends on the hostplant. The *Geum* sp. grow in humid and shady places, in woodlands and on mountain slopes, whereas the *Potentilla* sp. prefers dry, sandy and sunny localities.

REMARKS.– Adamczewski (1951) reports on breeding experiments with different foodplants, among them: *Leonurus* sp. and *Veronica officinalis* L. which had been reported as host-plants. The larvae refused the food and died and therefore these plant species are evidently not hostplants.

Procapperia Adamczewski, 1951

Procapperia Adamczewski, 1951: 338.
　Type species: *Oxyptilus maculatus* Constant, 1865; original designation.

DIAGNOSIS.– Palpi without a hair-brush along the third segment. The first forewing lobe acute, the second lobe terminally excavated (not as sinuate as in *Geina*). On the forewing lobes two white transverse lines. The scale-tooth at the dorsum of the third lobe of the hindwing well apart of the apex.

MALE GENITALIA.– Valvae symmetrical, simple, the distal half wider than the basal half. Aedeagus S-shaped, without cornuti.

FEMALE GENITALIA.– Ostium small and shaped as a triangular sclerotised plate, progressing into the slender ductus bursae. Bursa copulatrix vesicular, without a signum.

DISTRIBUTION.– The genus is spread in mountainous areas of South-East France, the

Balkan countries and North Africa, and extends further into Asia Minor, India, Sri Lanka and Central Asia.

68. *Procapperia maculatus* (Constant, 1865)

Oxyptilus maculatus Constant, 1865: 193.   Pl. 8: 2. Figs 82, 220.

DIAGNOSIS.– Wingspan 17-24 mm. Colour yellow-brown, the transverse lines pure white and obvoius. Third lobe of the hindwing whitish, with a large scale-tooth. Differs from *P. croatica* Ad. (see next species) by the absence of a greyish tinge and the considerably larger size.

MALE GENITALIA.– See genus description. Differs from *P. croatica* Ad. by the more rounded apex of the valve.

FEMALE GENITALIA.– See genus description. The difference from *P. croatica* Ad. is the more rounded shape of the sclerotised part of the 7th tergite.

DISTRIBUTION.– Alps of France and Italy, Pyrenees.

BIOLOGY.– The moth flies from June to the beginning of August. The hostplant is *Scutellaria alpina* L. (Chrétien, 1922; Nel, 1986b; Bigot et al., 1990) flowering on mountain slopes at 1000-2000 m. The larva feeds in June. After emerging the moth flies around the hostplant at dusk.

69. *Procapperia croatica* Adamczewski, 1951

Procapperia croatica Adamczewski, 1951: 342.   Figs 83, 221.

DIAGNOSIS.– Wingspan 14-16 mm. Colour greyish-yellow. See *P. maculatus*.

MALE GENITALIA.– See *P. maculatus*.

FEMALE GENITALIA.– See *P. maculatus*.

DISTRIBUTION.– Known only from the type-series from southern Croatia.

BIOLOGY.– The moth flies in June and July. The hostplant and early stages are unknown.

Paracapperia Bigot & Picard, 1986

Paracapperia Bigot & Picard, 1986b: [23].
 Type species: *Oxyptilus anatolicus* Caradja, 1920; original designation.

DIAGNOSIS.– Externally not differing from the genus *Capperia*.

MALE GENITALIA.– Valvae symmetrical, basally narrow, distally widened and club-like. Ninth sternite distally slighty bilobed, apical part densely setulose. Aedeagus weakly S-shaped, the apex narrow, tube-like.

FEMALE GENITALIA.– The ostium has a semi-circular shape, the ductus bursae ending centri-basal. Lamina antevaginalis rounded, with a large, elongated sclerotised margin.

DISTRIBUTION.– Palaeartic region.

70. *Paracapperia anatolicus* (Caradja, 1920)

Oxyptilus anatolicus Caradja, 1920: 79.   Pl. 8: 3. Figs 84, 222.
Capperia tamsi Adamczewski, 1951: 368.

DIAGNOSIS.– Wingspan 16-18 mm. Colour dark brown with a reddish to greyish gloss. On the dorsum of the second lobe of the hindwing a white spot in the middle of the cilia.

MALE GENITALIA.– See genus characters.

FEMALE GENITALIA.– See genus characters.

DISTRIBUTION.– Spain, Asia Minor and Syria.

BIOLOGY.– The moth flies in February (Spanish specimen in BMNH), and July and August (material from Turkey). The hostplant is unknown.

REMARKS.– The occurrence in Spain (Andalusia) seems doubtful. Bigot & Picard (1986) suggested that mis-labelling may have happened.

Capperia Tutt, 1905

Capperia Tutt, 1905: 37.
    Type species: *Phalaena heterodactyla* sensu Villers, 1789; misidentification; =*Oxyptilus britanniodactylus* Gregson, 1869; original designation.

DIAGNOSIS.– Palpi without a hair-brush along the third segment. No abdominal hair-brushes. First forewing lobe acute, second lobe with an excavated (not sinuate) terminal margin. Across the lobes two transverse white lines. At the third lobe of the hindwing a (sub)apical scale-tooth.

MALE GENITALIA.– Valvae strongly sclerotised, with processes and/or spines. Aedeagus S-shaped, occasionally with processes, ridges or plates. Ninth sternum sclerotised and extended as a bifurcated plate.

FEMALE GENITALIA.– Ostium in shape of an irregular plate, triangular or shield-like. Ductus bursae slender. Bursa copulatrix without a signum. Lamina antevaginalis well developed, often with a small sclerotised central plate.

DISTRIBUTION.– Holarctic.

BIOLOGY.– The species of this genus seem to be monophagous or oligophagous on Lamiaceae.

REMARKS.– The species of *Capperia* are very difficult to separate on external characters, and consequently a lot of confusion exists in the published records. The species treated below are characterized mainly by their genitalia, and external features are hardly used. In this way, one is forced to use the more troublesome way of identification by genital examination. I hope the result in the future will be a better reflection of the distribution of the species than provided in the old literature.

71. *Capperia britanniodactylus* (Gregson, 1867)

Oxyptilus britanniodactylus Gregson, 1867: 305.   Pl. 8: 4; pl. 16: 1-3. Figs 85, 223.
Pterophorus teucrii Jordan, 1869: 14.

DIAGNOSIS.– Wingspan 18-21 mm. Colour dark brown, with pure white markings. One of the darkest species, very similar to *C. fusca* which is smaller (14-16 mm).

MALE GENITALIA.– Valvae almost parallel-margined. Aedeagus strongly curved in an S shape, bifurcate near the tip. Two distal processes of uneven length, the longest of them half as long as the terminal process.

FEMALE GENITALIA.– Ostium in shape of an asymmetrical U. The ductus bursae ends in the left lateral part of the ostium. The lamina antevaginalis with a large triangular central plate, with curved base and apex.

DISTRIBUTION.– Western and central-western Europe.

BIOLOGY.– The moth flies from the end of May to the beginning of August, but mainly in June and the beginning of July. The hostplant is *Teucrium scorodonia* L. (Beirne, 1954; Hannemann, 1977b; Emmet, 1979; Bigot & Picard, 1986b; Nel, 1988c; Gielis, bred). The larva bites out a part of the stem below a group of well developed leaves, causing these to wither and later to drop. The larva lives in these leaves. Pupation takes place on the stem near the place where this has been damaged.

72. *Capperia celeusi* (Frey, 1886)

Oxyptilus celeusi Frey, 1886: 18.   Pl. 8: 5. Figs 86, 224
Oxyptilus intercisus Meyrick, 1930: 565.

DIAGNOSIS.– Wingspan 16-20 mm. Colour variable from dark brown, to yellow-brown and grey-brown.

MALE GENITALIA.– The margins of the valvae nearly parallel, apex rounded. Aedeagus strongly curved in an S-shape, with processes near the tip.

FEMALE GENITALIA.– Ostium asymmetrical U-shaped, "bottom" part curved to the right side. The ductus bursae centrally ending in the open part of the U. Lamina antevaginalis with a central trapezoid plate.

DISTRIBUTION.– Central Europe and the northern parts of ex-Yugoslavia.

BIOLOGY.– The moth flies from the end of May to the middle of June and again from the middle of July to the beginning of August. The hostplant is *Teucrium chamaedrys* L. (Frey, 1886; Adamczewski, 1951; Gozmány, 1962; Hannemann, 1977b; Nel, 1986b).

REMARKS.– This species has in most literature been cited with Schmid as the author. However Frey is the sole author of the paper including the description.

73. *Capperia trichodactyla* ([Denis & Schiffermüller], 1775)

Alucita trichodactyla [Denis & Schiffermüller], 1775: 145.   Pl. 8: 6. Figs 87, 225.
Oxyptilus leonuri Stange, 1882: 514.
Oxyptilus affinis Müller-Rutz, 1934: 118.

DIAGNOSIS.– Wingspan 15-20 mm. Colour dark brown, with a green-brown gloss. The scale-tooth at the end of the third lobe of the hindwing rounded.

MALE GENITALIA.– The margins of the valvae almost parallel, a basally directed large flap at one third and a weak, triangular process just before the end of the valvae. Aedeagus weakly S-curved; the apex thickened and cleft.

FEMALE GENITALIA.– Ostium almost circular with a smaller semi-circular sclerotised ridge. The ductus bursae ending on the left lateral side of the sclerotised ridge. Lamina antevaginalis with a trapezoid central plate, markedly wider at the base than on top part.

DISTRIBUTION.– Central Europe (recorded as far north as Finland) and South Yugoslavia.

BIOLOGY.– The moth flies in June and later in July and August. The hostplant is *Leonurus cardiaca* L. (Mitterberger, 1912; Schille, 1912; Adamczewski, 1948; Gozmány, 1962; Hannemann, 1977b; Buszko, 1979; Bigot & Picard, 1986b). The first generation larva eats the fresh top leaves, occasionally withered leaves, eating holes in the leaves. It hides in daytime, and only feeds if there is no wind, sun or rain. Pupation occurs on flower-stalks or on the main stem. Pupal stage lasts approximately ten days. The eggs of the second generation are deposited on flower-buds. These larvae feed on the flowers, causing them to drop.

REMARKS.– Adamczewski (1951) mentions the presence of lime soil where this species occurs.

74. *Capperia fusca* (Hofmann, 1898)

Oxyptilus leonuri Stange var. *fusca* Hofmann, 1898a: 339.   Pl. 8: 7. Figs 88, 226.
Capperia fusca Hofmann f. *marrubii* Adamczewski, 1951: 365.

DIAGNOSIS. – Wingspan 13-15 mm. Colour dark chocolate brown, with a reddish tinge. The white transverse markings reduced, almost absent.

MALE GENITALIA.– Valvae at middle conically widened. Aedeagus strongly S-shaped; the apex weakly bilobed.

FEMALE GENITALIA.– Ostium rounded, plate-like. The ductus bursae ending centrally (in contrast to *C. trichodactyla* which has a lateral genital pore). The lamina antevaginalis with a trapezoid central plate of a more rectangular shape than in *C. trichodactyla*.

DISTRIBUTION.– Western and Central Europe, especially in mountainous areas. The occurrence in the Balkans needs confirmation.

BIOLOGY.– The moth flies in June and from mid-July to mid-August. The hostplant is *Stachys alpina* L. (Hofmann, 1898a; Mitterberger, 1912; Adamczewski, 1951; Hannemann, 1977b; Buszko, 1986; Bigot & Picard, 1986b), and Buszko (1979) mentions

S. *cassia* (Boiss.) as an additional hostplant in Bulgaria. The larva of the first generation feeds on the leaves and stalks and pupates on the stem or under a leaf; the summer generation prefers the flower-buds and flowers and pupation follows near the feeding locality in a calyx. The form *marrubii* is recorded from *Marrubium vulgare* L. (Adamczewski, 1951; Hannemann, 1977b).

75. *Capperia bonneaui* Bigot, 1987

Capperia bonneaui Bigot, 1987: [35].　　　　　　　　　　　Pl. 8: 8. Figs 89, 227.

DIAGNOSIS.– Wingspan 16 mm. External characters as in the genus.

MALE GENITALIA.– Like *C. fusca*, but differs in the slightly wider and thicker aedeagus, which has a stretched S-shape. Distally it is bifid, with the apices close together. The ninth tergite has a conical tip, in contrast to the elongate pointed one seen in *C. fusca*.

FEMALE GENITALIA.– Ostium wide, saucer-like, covered with a longitudinally extended shield. Laterally this shield shows attached small hooks. The ductus bursae slender. Bursa copulatrix vesicular, covered in the apical half with minute spiculae.

DISTRIBUTION.– Spain, in the Valdevecar near Albarracin (province of Teruel).

BIOLOGY.– The moth flies in June and July. The hostplant is unknown.

76. *Capperia hellenica* Adamczewski, 1951

Capperia hellenica Adamczewski, 1951: 371.　　　　　　　Pl. 8: 9. Figs 90, 228.

DIAGNOSIS.– Wingspan 10-17 mm. Colour yellow-brown to brown. The forewing lobes slender.

MALE GENITALIA.– Valvae strongly arched with a rounded apex. The aedeagus is S-shaped, without processes or spines; the basal part twice as thick as the apical part.

FEMALE GENITALIA.– Ostium a small circular plate. The ductus bursae ending centrally in the ostium. Lamina antevaginalis rather acute, with a sclerotised margin extending beyond the ostium.

DISTRIBUTION.– Mediterranean area, from Spain to Asia Minor.

BIOLOGY.– The moth flies in July and August. The hostplant is *Stachys recta* L. (Nel & Prola, 1989; Bigot et al., 1990).

REMARKS.– Adamczewski (1951b) mentions a spring generation, but I have not seen specimens dated as such.

77. *Capperia loranus* (Fuchs, 1895)

Oxyptilus loranus Fuchs, 1895: 48.　　　　　　　　　　　Pl. 8: 10. Figs 91, 229.
Capperia sequanensis Gibeaux, 1990c: 73. **Syn. n.**

DIAGNOSIS.– Wingspan 15-17 mm. Colour brown-grey, transverse lines yellowish.

MALE GENITALIA.– Valvae with a long narrow and rounded basal flap projecting to the base. Aedeagus S-curved, bidentate and with a well developed terminal plate.

FEMALE GENITALIA.– Ostium large, shield-like. The opening of the ductus bursae in the distal, central part of the ostium. Lamina antevaginalis elongated, with a narrow sclerotised margin, reaching as far back as to the lateral margin of the ostium-plate.

DISTRIBUTION.– Belgium, France, the Rhineland (Germany), Austria and Slovakia.

BIOLOGY.– The moth flies at the end of May and the beginning of June and again in July and August. The supposed hostplant is *Teucrium botrys* L. (Arenberger, 1990b).

REMARKS.– The illustrated male genitalia of *C. sequanensis* are in the shape of the aedeagus, vinculum and tegumen identical to specimens of *loranus* examined. I consider the two identical and to be synonymised.

78. *Capperia marginellus* (Zeller, 1847)

Pterophorus marginellus Zeller, 1847: 903.   Pl. 8: 11. Figs 92, 230.

DIAGNOSIS.– Wingspan 15-17 mm. Colour dark chocolate-brown, with weak transverse markings.

MALE GENITALIA.– Valvae nearly straight and the margins almost parallel. A basal pointed flap present. Aedeagus strongly S-curved. The apex is bidentate, and the terminal part consists of a sclerotised, tridentate asymmetrical plate.

FEMALE GENITALIA.– Ostium shield-like, as in *C. loranus*, but the proximal margin slightly excavated, and wider. The ductus bursae ending in the centro-terminal part of the ostium and surrounded by a semi-circular, weakly sclerotised plate. The lamina antevaginalis with a well developed sclerotised ridge along the margin, beyond the widest part of the ostium.

DISTRIBUTION.– Sicily.

BIOLOGY.– The moth flies in May. The hostplant is unknown.

79. *Capperia zelleri* Adamczewski, 1951

Capperia zelleri Adamczewski, 1951: 375.   Pl. 9: 1. Fig. 93.

DIAGNOSIS.– Wingspan 14 mm. Colour pale brown, with a yellowish tinge.

MALE GENITALIA.– Valvae arched, distally almost twice as wide as near middle. The saccular flap pointing basally is long. Aedeagus strongly S-curved, with an asymmetrical terminal plate, margined with numerous small spines.

FEMALE GENITALIA.– Unknown.

DISTRIBUTION.– Sicily.

BIOLOGY.– The moth flies in July. The hostplant is unknown.

80. *Capperia polonica* Adamczewski, 1951

Capperia polonica Adamczewski, 1951: 376.   Pl. 9: 2. Figs 94, 231.

DIAGNOSIS.– Wingspan 14-18 mm. Colour dark brown, with strongly developed white markings on the wing and in the cilia.

MALE GENITALIA.– Valvae terminally half as wide as medially; the middle saccular process pointed. The strongly curved, S-shaped aedeagus has a bidentate apex, which has an asymmetrical dentated terminal plate, bearing small lateral spines.

FEMALE GENITALIA.– Ostium in shape of a shield-like plate, with lateral rounded edges. The ductus bursae ending in the top-central part of the ostium, and is margined by a wavy sclerotised ridge. Lamina antevaginalis without a sclerotised central part.

DISTRIBUTION.– Sardinia, France, Greece, Asia Minor.

BIOLOGY.– The moth flies in June and July. The hostplant is *Teucrium flavum* L. (Bigot & Picard, 1986b; Nel, 1986b; Bigot et al., 1990).

81. *Capperia maratonica* Adamczewski, 1951

Capperia maratonica Adamczewski, 1951: 377.   Pl. 9: 3. Figs 95, 232.

DIAGNOSIS.– Wingspan 14-18 mm. Colour dark brown. The scale-tooth at the third lobe of the hindwing is in apical position, and not in subapical position as in *marginellus*.

MALE GENITALIA.– Valvae curved, and twice as wide distally as medially. The basal pointed saccular flap irregularly shaped, and covered with numerous spines. Aedeagus strongly S-curved; the terminal plate with tridentate apex.

FEMALE GENITALIA.– Ostium in shape of a shield-like plate, with the lateral edges forming a double sclerotised ridge. The basi-central part slightly flattened. The ductus bursae ending in the centro-distal margin. Lamina antevaginalis with a small central sclerotised tip.

DISTRIBUTION.– Mediterranean area.

BIOLOGY.– The moth flies from June to September, probably in two generations. The hostplant is *Teucrium scordium* L. ssp. *scordioides* (Schreber) Maire & Petitmengin (Buszko, 1979a; Bigot & Picard, 1986b; Bigot et al., 1986; Nel, 1986b). This plant grows in humid localities.

Buckleria Tutt, 1905

Buckleria Tutt, 1905: 37.
   Type species: *Pterophorus paludum* Zeller, 1841; original designation.

DIAGNOSIS.– The palpi slender, longer than the diameter of an eye, and without a hair-brush along the third segment. Forewing cleft from middle, both lobes with an acute apex. At the dorsum of the third lobe of the hindwing are isolated dark scales, but not in the shape of a scale-tooth.

MALE GENITALIA.– Uncus extremely reduced. Valvae slender, with a valvular lobe originating in the middle of the valva.

FEMALE GENITALIA.– Antrum tube-like, twice as long as wide. Bursa copulatrix without a signum.

DISTRIBUTION.– The genus has a single Palaearctic representative in the whole of the region and also in the Indian subcontinent. Recently a second species of this genus has been described in the Nearctic region.

BIOLOGY.– The two known species feed on *Drosera* spp. These plants grow in bogs and moorlands. Especially in industrialized countries these habitats are becoming sparse and the chances of finding the species are decreasing rapidly.

82. *Buckleria paludum* (Zeller, 1841)

Pterophorus paludum Zeller, 1841: 866.   Pl. 9: 4. Figs 96, 233.
Trichoptilus paludicola Fletcher, 1907: 20.
Pselnophorus dolichos Matsumura, 1931: 1056.

DIAGNOSIS.– Wingspan 11-13 mm. See genus description. The species differs from *Stangeia siceliota* in the darker colour, the sparse dark scales at the dorsum of the third lobe of the hindwing, and in the genital structure in both male and female.

MALE GENITALIA.– See genus description.

FEMALE GENITALIA.– See genus description.

DISTRIBUTION.– The species occurs in the whole of the area, except for the extreme north and south. Because of the biology the populations are rather scattered.

BIOLOGY.– The moth flies in two generations, May-June and July-August. The eggs are deposited on the underside of the leaves of *Drosera rotundifolia* L. (Chapman, 1906b; Champion, 1910; Jäckh, 1936; Beirne, 1954; Hannemann, 1977b; Gielis, bred), growing in moorlands and peat-bogs. The larva feeds on the leaves and petioles. Pupation takes place on the hostplant or a nearby object or plant.

Oxyptilus Zeller, 1841

Oxyptilus Zeller, 1841: 765.
Type species: *Pterophorus pilosellae* Zeller, 1841; subsequent designation (Tutt, 1905).

DIAGNOSIS.– The palpi upcurved, along the third segment an appressed or parallel hairbrush. The abdomen has a small hairbrush laterally on the 8th sternite. Forewing dark to shining brown; the first lobe acute and the second lobe with a sinuate terminal margin. Both lobes of forewing with two transverse white lines. The third lobe of the hindwing has an apical, sometimes subapical, black scale-tooth (in the following genus *Crombrugghia* the scale-tooth is in some distance from the apex).

MALE GENITALIA.– Valvae symmetrical, bilobed. The terminal lobe of the valvae is of variable size between species. Tegumen bilobed, symmetrical. Aedeagus slightly curved, spiculate at the distal end.

FEMALE GENITALIA.– Antrum funnel-like, small. Ductus bursae long and slender. Bursa copulatrix vesicular, with two small bean-shaped signa. Apophyses posteriores well developed. Apophyses anteriores absent.

DISTRIBUTION.– Mainly Palaearctic, with a few representatives in the Nearctic and Afrotropical regions.

REMARKS.– The present genus closely resembles *Crombrugghia*, differing in the apical position of the scale-tooth on the third hindwing lobe and in the structure of the female genitalia. In *Buckleria* Tutt, the lobes of the valvae are not positioned terminally, but in the middle.

83. *Oxyptilus pilosellae* (Zeller, 1841)

Pterophorus pilosellae Zeller, 1841: 789.   Pl. 9: 5. Figs 9, 97, 234.
Pterophorus pilosellae var. *bohemanni* Wallengren, 1862: 16.

DIAGNOSIS.– Wingspan 15-21 mm. Colour red-brown, with ochreous-white transverse markings. The scale-tooth at the third lobe of the hindwing subapical; along the costa with short scales, along the dorsum the scales are longer. The basally placed scales at the dorsum longer than the apical scales, giving it an oblique, almost triangular appearance.

The scale-tooth in *O. pilosellae* is oblique and narrow at its terminal end. In *O. chrysodactyla* the shape of the scale-tooth is more square, and well developed at the costa of the lobe. *O. parvidactyla* has a similar square shape of the scale-tooth, but it is less developed at the costa; and in *O. ericetorum* the shape is rounded, and it is as large at the costa as on the dorsum of the lobe.

MALE GENITALIA.– The valvular lobe two thirds as long as the valve. Tegumen lobes distally rounded. Apex of aedeagus with a row of small hooks.

FEMALE GENITALIA.– Antrum short, funnel-shaped, with gradually narrowing margins. Ostium oval, ventrally more strongly sclerotised than dorsally.

DISTRIBUTION.– The species is known from the whole area, except for the Arctic regions. Specimens have also been recorded from Asia Minor.

BIOLOGY.– The moth flies from May to August. Eggs are deposited on the underside of the leaves. The egg stage lasts about 10 days. The larva feeds in a loose spinning near the central shoot, later entering into the central rosette of *Hieracium pilosella* (L.) (Purdey, 1910; Mitterberger, 1912; Gozmány, 1962; Hannemann, 1977b; Buszko, 1986; Nel, 1988a). In this place it may pupate, or it pupates under a leaf. The pupal stage lasts approximately 14 days. The moth is locally common in open localities in woods, among bushes and heather, where the hostplant grows.

REMARKS.– The var. *bohemanni* appears to be a colour form and occurs among normal *O. pilosellae* populations. In the northern parts of the area it is especially found in sandy habitats. The form has been recorded as far south as the coastal dune area of the Netherlands.

84. *Oxyptilus chrysodactyla* ([Denis & Schiffermüller], 1775)

Alucita chrysodactyla [Denis & Schiffermüller], 1775: 320.   Pl. 9: 6. Figs 98, 235.
Pterophorus hieracii Zeller, 1841: 827.

DIAGNOSIS.– Wingspan 15-21 mm. Colour red-brown with shining white transverse markings. The apical scale-tooth at the third lobe of the hindwing has rather long scales at the costa which are almost parallel to the costa, and dorsally very long scales at right angles to the dorsum, giving the scale-tooth an almost rectangular, slightly backwardly pointed appearance.

MALE GENITALIA.– The valvular lobe 65-70 per cent as long as the valve. The distal part of the tegumen gradually narrowing, with almost rectangular angles. The apex of the aedeagus with a semi-circular row of small spines.

FEMALE GENITALIA.– Antrum short, gradually narrowing. The ostium hardly expressed. Margin of 7th tergite bisinuate, with a central excavation near the antrum.

DISTRIBUTION.– The species is known from the whole area, except for the far north and south. Its presence in England has not yet been established, but may be expected.

BIOLOGY.– The moth flies from June to August. Waste lands are preferred as well as lightly wooded areas. The larva feeds in the heart of *Hieracium umbellatum* L. (Mitterberger, 1912; Gozmány, 1962; Hannemann, 1977b; Aarvik et al., 1986; Buszko, 1986; Nel, 1988a), *H. amplexicaula* L. (Burmann, 1950, as *C. kollari*), *H. sabaudum* L. (=*H. boreale* Fries) (Bigot & Picard, 1988a; Nel, 1988a; Bigot et al., 1990) and *Picris hieracioides* L. (Mitterberger, 1912; Gozmány, 1962; Hannemann, 1977b; Buszko, 1986) spinning the top-leaves together. Pupation on the upper surface of a leaf or along the stem.

REMARKS.– The Denis & Schiffermüller collection was lost as a result of fire. Arenberger (1988a) published a neotype of *Alucita chrysodactyla* [Denis & Schiffermüller], and by this action fixed the name as a senior synonym of *Pterophorus hieracii* Zeller, 1841.

85. *Oxyptilus ericetorum* (Stainton, 1851)

Pterophorus ericetorum Stainton, 1851: 28.   Pl. 9: 7. Figs 99, 236.

DIAGNOSIS.– Wingspan 16-19 mm. Very closely related to *O. pilosellae*, but differing in the darker space between the faciae on the forewing and the more obliquely positioned white fasciae. The scale-tooth at the third lobe of the hindwing is apical. The shape of the scale-tooth is rounded, caused by the similar formation of the scales at the costa and the dorsum, i.e. the basal scales are large and become gradually shorter toward the apex.

MALE GENITALIA.– Valvular lobe seven tenths as long as the valve. The tegumen has the terminal margins parallel, and the distal corners are at right angles. Aedeagus with a single longitudinal row of spiculae near the apex.

FEMALE GENITALIA.– Antrum funnel-shaped, gradually narrowing. The ostium is weakly developed. The margin of the 7th tergite slightly excavated (unlike *O. chrysodactyla*).

DISTRIBUTION.– The species has been recorded from North and Central Europe, and in the cooler (mountainous) parts of the Mediterranean countries.

BIOLOGY.– The moth flies from June to August. As hostplants *Hieracium pilosella* (L.) (Mitterberger, 1912; Hannemann, 1977b; Buszko, 1986) and *H. murorum* L. (Nel, 1988a; Bigot et al., 1990) have been recorded. The larva feeds on the central shoot, the hairy surface of the leaves remaining as detritus around the area where the larva has fed. Pupation on a leaf or on the stem. Gozmány (1962) mentions *Calluna vulgaris* L. as a hostplant.

86. *Oxyptilus parvidactyla* (Haworth, 1811)

Alucita parvidactyla Haworth, 1811: 480.   Pl. 9: 8, 9. Figs 100, 237.
Pterophorus obscurus Zeller, 1841: 793.
Oxyptilus hoffmannseggi Möschler, 1866: 145.
Oxyptilus maroccanensis Amsel, 1956: 22.

DIAGNOSIS.– Wingspan 13-18 mm. Forewing colour dark to black-brown, transverse markings white. The scale-tooth at the apex of the third lobe of the hindwing is small at the costa and large and rectangular at the dorsum.

MALE GENITALIA.– Valvular lobe about one third as long as the valve. The tegumen margins are gradually converging towards the apex, and here curved outward. Aedeagus with club-like apex, which has a double row of spiculae.

FEMALE GENITALIA.– Antrum funnel-like, gradually narrowing, covered by a trapezoid sclerotised plate formed by the 7th tergite.

DISTRIBUTION.– Europe, Asia Minor and North Africa.

BIOLOGY.– The moth flies from May to July. The moth flies around the hostplant, *Hieracium pilosella* (L.) (Mitterberger, 1912; Gozmány, 1962; Hannemann, 1977b; Buszko, 1986; Nel, 1988a) and *H. laevigatum* group (Gozmány, 1962), in open spaces in woodlands and on heaths. Eggs are deposited on the underside of a leaf. The larva feeds on young leaves, later working its way into the central stem. Pupation on the surface of the plant.

REMARKS.– I have not examined the type specimen of the species described as *O. hoffmannseggi*. Prof. Dr H.-J. Hannemann informed me that most Möschler types were destroyed in the second world war, and so this type is probably lost. The description is however clear, mentioning the dark colour and the white markings on the forewing, together with the distinctly white third lobe of the hindwing. The genitalia of the male and female of a typical *O. parvidactyla* and *O. hoffmannseggi* are identical. The character of the white third lobe of the hindwing is, however, less reliable. In a series collected in the southern and central parts of Spain this character is present in a great majority of the specimens. In series collected in the north-eastern and northern parts of Spain, however, the number of "typical" *O. hoffmannseggi* specimens is low and most are typical *O. parvidactyla* specimens. This brings me to the conclusion that *O. hoffmannseggi* is a junior synonym of *O. parvidactyla*. In Spain, the species is expressed by: forma *hoffmannseggi*.

Crombrugghia Tutt, 1907

Crombrugghia Tutt, 1907: 449.
  Type species: *Pterophorus distans* Zeller, 1847; subsequent designation (Meyrick, 1910).
Combrugghia Neave, 1939; incorrect spelling.

DIAGNOSIS.– External characters as in *Oxyptilus*, except for the scale-tooth at the dorsum of the third lobe of the hindwing, which is subterminally to medially placed.

MALE GENITALIA.– As in *Oxyptilus* Zeller.

FEMALE GENITALIA.– As in *Oxyptilus* Zeller, but for the presence of a sclerotised plate beside the ostium and antrum part of the ductus bursae.

DISTRIBUTION.– Mainly in the western half of the Palaearctic region. According to this author, the species *wahlbergi* Zeller appears to belong to the genus. It is known from the eastern part of the Palaearctic region and Africa.

87. *Crombrugghia distans* (Zeller, 1847)

Pterophorus distans Zeller, 1847: 902.   Pl. 9: 10. Figs 10, 101, 238.
Oxyptilus clarisignis Meyrick, 1924: 93.
Oxyptilus buvati Bigot & Picard, 1988a: 242.
Oxyptilus buvati propedistans Bigot & Picard, 1988a: 244.
Oxyptilus adamczewskii Bigot & Picard, 1988a: 246.
Oxyptilis pravieli Bigot, Nel & Picard, 1989: 15.
Oxyptilus gibeauxi Bigot, Nel & Picard, 1990: 47.
Oxyptilus jaeckhi Bigot & Picard, 1991: 236.

DIAGNOSIS.– Wingspan 15-22 mm. Colour pale to dark red-brown. The scale-tooth on the third lobe of the hindwing at two thirds of the dorsum. The basal scales larger than the terminal scales, giving the scale-tooth a triangular appearance.
    The present species shows little difference with regard to external appearance and genital structure from the following two species: *C. tristis* and *C. kollari*. Although Bigot & Popescu-Gorj (1973) synonymized these species (later recalled again by Bigot), I prefer to treat them as separate, considering the constant differences. *C. tristis* is in general a smaller species and has a more brown-grey colour than the red-brown *C. distans*, and *C. kollari* is a remarkable grey-white species. The genitalia of *C. tristis* differ in the shape of the aedeagus and the female 7th tergite. In *C. kollari* the structures of the aedeagus and 7th tergite show further differences.

MALE GENITALIA.– Valvular lobe about nine tenths as long as the valve. The margins of tegumen gradually converging, and the edges rounded. The apex of the aedeagus club-like, with numerous fine spiculae internally.

FEMALE GENITALIA.– Antrum funnel-like, only slightly narrowing, laterally with longitudinal sclerotised lines, originating from the ostium. Lateral to the antrum, on both sides and well separated from each other, a semi-circular sclerotised plate.

DISTRIBUTION.– Europe except the northern parts, Asia Minor, North Africa, Canary Islands.

BIOLOGY.– The moth flies in two generations in April-June and July-September. Dry sandy waste-lands and park areas are preferred habitats. The larva of the spring generation feeds in the central parts of *Crepis capillaris* (L.) (Hannemann, 1977b; Emmet, 1979; Nel, 1988a) and *C. tectorum* L. (Hering, 1891; Hannemann, 1977b; Buszko, 1986; Nel, 1988a; Arenberger & Jaksic, 1991). The summer brood is recorded feeding on the flowers. Pupation on flowers, leaves or shoots. As additional host-plants are mentioned: *Hieracium pilosella* (L.) (Hannemann, 1977b; Emmet, 1979), *H. amplexicaule* L. (Bigot & Picard, 1988a; Nel, 1988a; Bigot & Picard, 1991), *Crepis succifolia* Tausch (Nel, 1988a), *C. conyzaefolia* (L.) Della Torre (Bigot & Picard, 1988a; Nel, 1988a; Bigot & Picard, 1991), *C. albida* Vill. (Bigot & Picard, 1991), *C. capillaris* L. (Wallr.) (Nel, 1988a; Bigot & Picard, 1991), *Sonchus asper* (L.) Hill. (Nel, 1988a; Bigot & Picard, 1991), *S. arvensis* L. (Nel, 1988a; Bigot & Picard, 1991), *Cichorium intybus* L. (Bigot & Picard, 1991) and *Picris hieracioides* L. (Hannemann, 1977b; Emmet, 1983; Bigot et al., 1990; Bigot & Picard, 1991). Pupal stage of approximately 10 days duration.

REMARKS.– The species described by Bigot, Nel & Picard show minor differences from *C. distans*. In their illustrations of the male genitalia, it is noticeable that the structures are illustrated inconsistantly. If prepared in a uniform way, the genital structures show no substantial difference. In the female genitalia the same remarks apply regarding the sclerotised plates lateral to the ostium. The argument that the species have different hostplants in the closely related genera of *Hieracium* and *Crepis*, are only valid if the concept of a strict monophagous life style is true. The differences indicated in the chaetation of the larvae are invalid as well, because the larvae examined are 6th instar larvae (Wasserthal, 1970). The differences in the head capsule markings are to be judged as intra-specific variation.

88. *Crombrugghia tristis* (Zeller, 1841)

Pterophorus tristis Zeller, 1841: 788.   Pl. 9: 11, 12. Figs 102, 239.

DIAGNOSIS.– Wingspan 14-19 mm. Colour brown-grey, the transverse markings and spots dark greyish. The scale-tooth at the third lobe of the hindwing as in *C. distans*.

MALE GENITALIA.– Valvular lobe and valve as in the preceding species. The tegumen lobes have more rounded apices. The aedeagus has a semicircular, lateral row of spiculae at the apex.

FEMALE GENITALIA.– The antrum margins more parallel than in *C. distans*, without lateral sclerotised lines. The sclerotised plates of the 7th tergite beside the antrum are of a more irregular shape than in *C. distans* and medially closer to each other.

DISTRIBUTION.– Central and southern Europe, especially in mountainous areas and the north-eastern part of Central Europe.

BIOLOGY.– The moth flies in two generations: June and August, in open localities in pine and birch woods, heaths and poor barren grasslands. The larva feeds on *Hieracium echioides* Lumnitzer (Hering, 1891; Mitterberger, 1912; Hannemann, 1977b), *H. umbelliferum* Nag. & Peter (Schwarz, 1953), *H. dubium* L. (LHomme, 1939), *H. cymosum* L. (Nel, 1988a), *H. piloselloides* Vill. (Nel, 1988a), *H. fallax* Willd. (Hering, 1891), *H. pilosella* (L.) (Hering, 1891; Hannemann, 1977b; Buszko, 1986) and *H.*

amplexicaule L. (Burmann, 1986), in the central parts or near the rosette-leaves and are hidden by the masses of felt-like hairs of the leaves. Pupation in general on the upper surface of a rosette-leaf.

89. *Crombrugghia kollari* (Stainton, 1851)

Pterophorus kollari Stainton, 1851: 28.   Pl. 10: 1. Figs 103, 240.

DIAGNOSIS.– Wingspan 16-21 mm. Colour of the forewing grey-white, in some specimens almost white.

MALE GENITALIA.– Valvular lobe about five sevenths as long as the valve. The margins of the tegumen lobes gradually converging and ending in a rounded apex. Aedeagus with a semi-circular row of spiculae near the apex.

FEMALE GENITALIA.– Antrum funnel-like, gradually narrowing, without lateral sclerotised lines. The lateral sclerotised plates semi-circular and medially fused.

DISTRIBUTION.– Alpine regions in Central Europe: France, Austria, Switzerland. The record for France is based on a single specimen bearing the label: "Gallia". Zagulajev (1986) mentions the species from Asia.

BIOLOGY.– The moth flies in July and August. Hostplant and early stages unkown. The species described from *Hieracium amplexicaula* L. by Burmann (1950) is in fact *O. chrysodactyla.*

90. *Crombrugghia laetus* (Zeller, 1847)

Pterophorus laetus Zeller, 1847: 903.   Pl. 10: 2. Figs 104, 241.
Pterophorus loetidactylus Bruand, 1859: 893.
Oxyptilus lantoscanus Millière, 1883: 176.

DIAGNOSIS.– Wingspan 14-23 mm. Colour brown-ochreous to pale brown, differing in general from *C. distans* by the more ochreous tinge. The scale-tooth on the third lobe of the hindwing is less well developed than in *C. distans.*

MALE GENITALIA.– Valvular lobe about one third as long as the valve. Tegumen arms slightly irregularly shaped before the apex. Aedeagus with a semi-circular row of spiculae at the apex.

FEMALE GENITALIA.– Antrum tubular, as long as wide. The ostium roundish, with a ventro-caudal bulge. The 7th tergite with two bulges, the central space excavated at the antrum. Ventro-rostral of the antrum with two sclerotised small plates.

DISTRIBUTION.– Southern Europe, almost restricted to the Mediterranean area. The species also occurs in North Africa, the Canary Islands, Asia Minor and Iraq.

BIOLOGY.– The moth flies from May to October, in several successive generations. The habitats are dry and hot grasslands and shrubby areas. The hostplant is *Andryale integrifolia* L. (=*sinuata* L.) (Nel, 1988a; Bigot et al., 1990). In case of the synonymized *C. lantoscanus, Hieracium lanatum* Vill. (Bigot & Picard, 1981; Nel, 1988a) is mentioned as a foodplant.

REMARKS.– I have compared the type specimens of *C. laetus* and *C. lantoscanus*. Differences are found in the larger size and brighter colour of the latter, but the genitalia are identical. I therefore conclude that these species are synonyms. See remarks under *O. pilosellae* var. *bohemanni*.

Stangeia Tutt, 1905

Stangeia Tutt, 1905: 37.
  Type species: *Pterophorus siceliota* Zeller, 1847; monotypy.

DIAGNOSIS.– Forewing with the apex of both lobes acute. At the dorsum of the third lobe of the hindwing a row of black and white scales, and an ill-defined scale-tooth at two thirds of distance from base.

MALE GENITALIA.– Valvae very slender, simple. Tegumen and vinculum well developed. Aedeagus conical, almost straight.

FEMALE GENITALIA.– Ostium in shape of a rectangular, oblique plate. Antrum rounded into this plate, short. Ductus bursae long and slender. Bursa copulatrix vesicular, covered with numerous spiculae. Lateral to the ostium plate, the 7th tergite expands into a pair of longitudinal lobes, as long as the diameter of the ostium plate.

DISTRIBUTION.– The genus is represented by *S. siceliota* in our area. One representative is known from Africa; and in the Indo-Australian and Pacific regions three more species are recognized.

BIOLOGY.– See *S. siceliota*.

91. Stangeia siceliota (Zeller, 1847)

Pterophorus siceliota Zeller, 1847: 907.   Pl. 10: 3. Figs 105, 242.
Pterophorus ononidis Zeller, 1852: 401.

DIAGNOSIS.– Wingspan 12-16 mm. See genus description.

MALE GENITALIA.– See genus description.

FEMALE GENITALIA.– See genus description.

DISTRIBUTION.– The Mediterranean area, extending through Asia Minor into Iraq. To the north the species has been recorded from southern Switzerland, Valais.

BIOLOGY.– The moth flies from April to October. The hostplant is *Cistus monspeliensis* L. (Nel, 1986c; Arenberger & Jaksic, 1991). Other recorded hostplants are *C. albidus* L. (Nel, 1986c), and *C. salviaefolius* L. (Nel, 1986c), *Sanguisorba* (=*Poterium*) sp., *Dittrichia viscosa* L. (Nel, 1986c; Arenberger & Jaksic, 1991) and *Ononis natrix* L.

Megalorhipida Amsel, 1935

Megalorhipida Amsel, 1935a: 283.
  Type species: *Megalorhipida palaestinensis* Amsel, 1935; monotypy.

DIAGNOSIS.– Forewing with faint transverse whitish lines on the lobes. Wing venation of forewing: Sc to costa, R1 short, R2 and R3 fused, R4 and R5 fused to apex of first lobe. M and Cu fused to apex of second lobe, An to half the wing-length and Ax short. Hindwing with a small medially placed scale-tooth at the dorsum of the third lobe. Venation of hindwing: Sc, RR and M1 fused, Cu1 to apex of second lobe, Cu2 to half the second lobe and An as far as base of first cleft, An to apex of third lobe.

MALE GENITALIA.– Valvae symmetrical, rounded. Tegumen with a swollen, hairy uncus. Aedeagus almost straight, without cornuti.

FEMALE GENITALIA.– Bursa copulatrix with a pair of signa. Ductus bursae narrow, tube-like. Antrum flattened. Margin of 7th sternite excavated. Mid-section of 8th sternite posteriorly extended. Apophyses anteriores absent. Apophyses posteriores 3 to 4 times as long as papillae anales.

DISTRIBUTION.– Pantropical and subtropical.

BIOLOGY.– The recorded hostplants are *Acacia neovernicosa* Isely (*C. vernicosa* Standl.) (Fabaceae), *Boerhavia diffusa* L. (Nyctaginaceae), *Okenia hypogae* Schlecht. and Cham. (Nyctaginaceae), *Amaranthus* (Amaranthaceae) and *Mimosa* (Mimosaceae) (Wolcott, 1936; Zimmerman, 1958; Matthews, 1989).

REMARKS.– The genus has a (sub)tropical distribution. In our area it reaches its northern limit in the south of Spain.

92. *Megalorhipida leucodactylus* (Fabricius, 1794), comb. nov.

Pterophorus leucodactylus Fabricius, 1794: 346.   Pl. 10: 4. Figs 106, 243.
Pterophorus defectalis Walker, 1864: 943.
Pterophorus congrualis Walker, 1864: 943-944.
Pterophorus oxydactylus Walker, 1864: 944.
Trichoptilus ochrodactylus Fish, 1881: 142.
Aciptilia hawaiiensis Butler, 1882: 408.
Trichoptilus centetes Meyrick, 1886: 16.
Trichoptilus compsochares Meyrick, 1886: 16.
Trichoptilus ralumensis Pagenstecher, 1900: 239.
Trichoptilus derelictus Meyrick, 1926: 276.
Megalorhipida palaestinensis Amsel, 1935a: 293.

DIAGNOSIS.– Wingspan 14-17 mm. The species is characterized by the medially placed scale-tooth on the third lobe of the hindwing.

MALE GENITALIA.– See genus description.

FEMALE GENITALIA.– See genus description.

DISTRIBUTION.– Southern Spain and Israel. The species has a circum-tropical distribution, extending north as far as the southern margins of the area treated.

BIOLOGY.– The moth flies in April and May and in July. The hostplant is *Boerhavia repens* L.; the larva feeds on the unripe seeds (Fletcher, 1909; Fletcher, 1921). Other hostplants are mentioned under the genus description.

VARIATION.– In the specimens from the Caribbean area the androconial scales are dark brown. In general the specimens from the Americas have a paler colour than those from Africa. The specimens examined from the Galapagos Archipelago and Socorro Island, Mexico show a general darkening of the colour.

REMARKS.– The synonymy of *leucodactylus* and *defectalis* is discussed by Karsholt & Gielis (1995).

Puerphorus Arenberger, 1990

Puerphorus Arenberger, 1990a: 18.
  Type species: *Pterophorus olbiadactylus* Millière, 1859; original designation.

DIAGNOSIS.– Forewing cleft starts before middle. The veins R1 and R5 absent. In the second lobe both Cu-veins present. In the hindwing each lobe has two veins.

MALE GENITALIA.– Valvae strongly asymmetrical. Left valve broadly rounded apically, and with a sacculus having two medially directed protrusions. Right valve with an acute apex, and a sacculus with a single protrusion. Uncus bifurcate. Vinculum rounded. Saccus wide and apically provided with a hair-tuft. Aedeagus with lateral spines near the apex.

FEMALE GENITALIA.– A large oval ostium plate, with laterally ending antrum. Ductus bursae and bursa copulatrix simple. Apophyses anteriores absent.

DISTRIBUTION.– Mediterranean area.

BIOLOGY.– The hostplants are two *Phagnalon* spp. (Asteraceae).

93. *Puerphorus olbiadactylus* (Millière, 1859)

Pterophorus olbiadactylus Millière, 1859: 89.   Pl. 10: 5. Figs 107, 244.
Pselnophorus hemiargus Meyrick, 1907: 491.
Gypsochares dactilographa Turati, 1927: 335.

DIAGNOSIS.– Wingspan 14-17 mm. Colour white-ochreous. The second lobe of forewing almost completely white. Along the dorsum of the second lobe a row of well developed ochreous grey scales. The species can easily be mistaken for a *Pterophorus*. It is best recognized by the well developed scales on the dorsum of the second forewing lobe.

MALE GENITALIA.– See genus description.

FEMALE GENITALIA.– See genus description.

DISTRIBUTION.– Mediterranean area and Canary Islands.

BIOLOGY.– The moth flies in January and from March to May. The hostplants are *Phagnalon rupestre* L. (LHomme, 1939) and *P. saxatile* L. (Nel, 1987c; Nel, 1989c). The larva eats the parenchyma from the upperside, causing the felted underside to curl around.

Gypsochares Meyrick, 1890

Gypsochares Meyrick, 1890: 488.
  Type species: *Pterophorus baptodactylus* Zeller, 1850; monotypy.

DIAGNOSIS.– Forewing cleft from just beyond middle. Vein R5 close to the base of R4.

Male genitalia.– Valvae asymmetrical. Both valvae have saccular processes, which may be forked.

FEMALE GENITALIA.– Antrum well developed, widely funnel-shaped.

DISTRIBUTION.– Mediterranean area, and extending into southern Asia and the Himalaya Mountains.

BIOLOGY.– The larva feeds on hostplants of the genus *Helichrysum* Miller (Asteraceae).

94. *Gypsochares baptodactylus* (Zeller, 1850)

Pterophorus baptodactylus Zeller, 1850: 211.   Pl. 10: 6. Figs 108, 245.

DIAGNOSIS.– Wingspan 11-17 mm. Colour brown-yellow. The second lobe white. Markings black-brown, consisting of two costal spots in the first lobe, a spot at the base of the cleft, and a small discal spot.

MALE GENITALIA.– Left valve angulated, with a long and wide cucullar spine, extending beyond the apex of the valve. Right valve more rounded, with a cucullar process beyond the apex of the valve. Saccus with a single apex.

FEMALE GENITALIA.– Ostium smoothly excavated. Antrum wide, bell-shaped. Apophyses anteriores short.

DISTRIBUTION.– Mediterranean area.

BIOLOGY.– The moth flies from May to September. The hostplants are *Helichrysum italicum* Roth (Bigot & Picard, 1983; Gibeaux & Nel, 1990a).

REMARKS.– The distribution of this and the next species is uncertain, because of their external resemblance. As they seem to be at least partly sympatric, all old distribution data need to be verified.

95. *Gypsochares bigoti* Gibeaux & Nel, 1990

Gypsochares bigoti Gibeaux & Nel, 1990a: 121.   Pl. 10: 7. Figs 109, 246.

DIAGNOSIS.– Wingspan 15-17 mm. External characters as in *G. baptodactyla*.

MALE GENITALIA.– Left valve more rounded than in previous species. The cucullar spine is longer and more curved apically. The right valve has a more vesicular top than the previous species, and the cucullar spine is better developed. The saccus is bidentate.

FEMALE GENITALIA.– Ostium deeply excavated. Antrum small and hardly narrowing. Apophyses anteriores long.

DISTRIBUTION.– Southern France.

BIOLOGY.– The moth flies in May. The hostplant is *Helichrysum stoechas* L. (Bigot et al., 1990; Gibeaux & Nel, 1990a).

96. *Gypsochares nielswolffi* Gielis & Arenberger, 1992

Gypsochares nielswolffi Gielis & Arenberger, 1992: 81.   Pl. 10: 8. Fig. 110.

DIAGNOSIS.– Wingspan 15-18 mm. The species is characterized by the brown colour of the forewing, with a distinct grey-white mid-costal line and second forewing lobe. The related *G. baptodactyla* and *G. bigoti* have only limited white scaling along the costa between the dark costal spots in the first lobe.

MALE GENITALIA.– Valvae asymmetrical. Left valve with a ventro-basal knob-like cucullar process and a stout, hooked, and curved ventral process. The right valve with a spine-like cucullar apex and a rather small, ventral cucullar process. The vinculum wide and arched, with a single saccus protrusion. Tegumen and uncus simple. The juxta extended into a single ventrally curved process, which is as long as the uncus. Aedeagus slightly curved.

FEMALE GENITALIA.– Unknown.

DISTRIBUTION.– Madeira.

BIOLOGY.– The moth flies in August and September. The hostplant is unknown.

REMARKS.– The species seems to be the only endemic species of plume moth on Madeira.

Pselnophorus Wallengren, 1881

Pselnophorus Wallengren, 1881: 96.
  Type species: *Alucita brachydactyla* Kollar, 1832; monotypy.
Crasimetis Meyrick, 1890: 489.
  Type species: *Alucita brachydactyla* Kollar, 1832; subsequent designation (Meyrick, 1910).

DIAGNOSIS.– Palpi curved upwards, slender. Antennae densely ciliated. Hind legs with scale-brushes around the base of the spurs. Lateral spurs shorter than medial spurs. Forewing without anal angles in the lobes. Veins: R1 absent; R2 separate; R3-5 stalked; M3, Cu1 and Cu2 stalked. Hindwing vein M3 absent.

MALE GENITALIA.– Valvae asymmetrical, sacculus with well developed thorns or spines.

FEMALE GENITALIA.– Antrum and ductus bursae simple. Bursa copulatrix vesicular, signum absent or poorly developed.

DISTRIBUTION.– In the Palaearctic region three species are known. The species described in this genus from other faunal areas should be examined for their generic affinity.

BIOLOGY.– See *P. heterodactyla*.

REMARKS.– The genus is closely related to the *Pterophorus* group of genera and the *Oidaematophorus* group, differing mainly in the wing venation.

97. *Pselnophorus heterodactyla* (Müller, 1764)

Phalaena Alucita heterodactyla Müller, 1764: 59. Pl. 10: 9. Figs 4, 111, 247.
Alucita brachydactyla Kollar, 1832: 100.
Pterophorus aetodactylus Duponchel [*in*: Godart], 1840a: 659.

DIAGNOSIS.– Wingspan 18-22 mm. Colour black-brown to dark brown. White costal spots near the discal area; at the base of the cleft, two small spots at the first lobe and subapical; a white spot around the base of the cleft; and an indistinct discal spot. Fringes dark brown, with two white hairstreaks at the dorsum of the second lobe.

MALE GENITALIA.– Left valve wider and more rounded than the right valve, and with a short and stout, curved saccular spine. The right valve with a more slender, longer and less curved spine.

FEMALE GENITALIA.– Ostium one and a half times as wide as the antrum, flattened. Antrum funnel-shaped, strongly and suddenly curved and narrowing, before progressing into the membraneous ductus bursae. Bursa copulatrix vesicular, with fine, delicate spiculation.

DISTRIBUTION.– Widespread in the area. Not yet recorded from Spain or Norway. In Central Europe more common, but local.

BIOLOGY.– The moth flies from June to mid August. The hostplants are *Mycelis muralis* L. (Mitterberger, 1912; Sheldon, 1932; Beirne, 1954; Gozmány, 1962; Hannemann, 1977b; Buszko, 1986), *Prenanthes purpurea* L. (Mitterberger, 1912; Gozmány, 1962; Hannemann, 1977b; Buszko, 1986) and *Lapsana communis* L. (Mitterberger, 1912; Gozmány, 1962; Hannemann, 1977b; Buszko, 1986; Bigot et al., 1990). The hostplants grow in (mountainous) woodlands with a herbaceous ground cover. The larva feeds on the underside of a leaf. In August the feeding is stopped and hibernation takes place in a withered leaf. In spring the larva recommences feeding on a leaf, biting through its midrib and causing the tip to droop. Pupation along the stem of the hostplant.

Hellinsia Tutt, 1905

Hellinsia Tutt, 1905: 37.
    Type species: *Pterophorus osteodactylus* Zeller, 1841; original designation.
Leioptilus Wallengren, 1862: 21; homonym.
    Type species: *Alucita tephradactyla* Hübner, [1813]; subsequent designation (Tutt, 1905).
Lioptilus Zeller, 1867; emendation.

DIAGNOSIS.– Forewing cleft for apical third, the first forewing lobe without an anal angle. Forewing veins: R1 from the costa of the discal cell, R2+3, R4 and R5 separate near the apical corner of the discal cell. Middle legs with an ill-developed scale-brush around the base of the spur pairs.

MALE GENITALIA.– Valvae asymmetrical. Sacculus in the left valve with a well developed spine, in the right valve occasionally with small thorns or spines.

FEMALE GENITALIA.– Antrum in general laterally placed, with limited sclerotised parts. Bursa copulatrix vesicular, often well developed, rarely with a signum. Vesica seminalis mostly well developed.

DISTRIBUTION.– Known from all faunistic regions, with the largest number of species in the New World.

BIOLOGY.– The hostplants belong to the Asteraceae.

98. *Hellinsia inulae* (Zeller, 1852)

Pterophorus inulae Zeller, 1852: 384.
Pterophorus coniodactylus Staudinger, 1859: 258.

Pl. 10: 10. Figs 112, 248.

DIAGNOSIS.– Wingspan 16-20 mm. Colour brown-yellow, densely speckled with brown scales. A spot before the base of the cleft; beyond this a costal spot and small spots at the dorsum of both lobes. *H. inulae* resembles *H. carphodactyla* very much, but differs in the more dull brown-yellow colour. The status of the two taxa is not yet clear.

MALE GENITALIA.– Left valve more elongate than in *H. carphodactyla*. At the sacculus of the right valve two small spines (in *H. carphodactyla* one). The aedeagus has a cornutus, that is slightly larger than in *H. carphodactyla*.

FEMALE GENITALIA.– The genital structure resembles that of *H. carphodactyla*, but differs in the wider funnel-shaped antrum and the apophyses anteriores which are straight and not bifurcate.

DISTRIBUTION.– Central Europe, Mediterranean area and Canary Islands. Extends into North Africa and Asia Minor.

BIOLOGY.– The moth flies in Central Europe from July to September. In the Mediterranean area from March to October. The hostplants are *Inula britannica* L. (Mitterberger, 1912; Gozmány, 1962; Hannemann, 1977b; Buszko, 1986; Gibeaux & Picard, 1992), *I. salicina* L. (Mitterberger, 1912; Gozmány, 1962; Gibeaux & Picard, 1992), *I. viscosa* (L.) Aït. (Gibeaux & Picard, 1992) and *Dittrichia viscosa* Aït. These plants prefer wet places like marshes, wet meadows, the sides of ditches, brooks and ponds. The larva feeds on the flower-buds causing brown spotting of the plant parts. Later on it consumes unripe seeds and eventually it may enter the stem. Pupation in the withered flower-head. The pupal period lasts 2 weeks.

99. *Hellinsia carphodactyla* (Hübner, [1813])

Alucita carphodactyla Hübner, [1813]: t. 4, figs 19, 20.
Leioptilus carphodactylus var. *buphthalmi* Hofmann, 1898a: 340.
Leioptilus inulaevorus Gibeaux, 1989d: 73.
Oidaematophorus alpinus Gibeaux & Picard, 1992: 123. **Syn. n.**

Pl. 10: 11. Figs 113, 249.

DIAGNOSIS.– Wingspan 14-23 mm. Colour bright yellow. Dark brown spots at the base of the cleft; a costal spot on the first lobe obliquely beyond the base of the cleft; a small dorsal spot at the first lobe and three dorsal spots at the second lobe.

MALE GENITALIA.– Left valve rounded, with an acute, broad-based, saccular spine. Right valve narrower, with a single small saccular spine. Aedeagus with a small cornutus.

FEMALE GENITALIA.– Ostium and antrum small, funnel-shaped. Apophyses anteriores forked.

DISTRIBUTION.– Europe, except for Scandinavia. Extending into Asia Minor and North Africa.

BIOLOGY.– The moth flies from May to July, and in August and September. The host-plants are *Inula conyza* C.D. (Schmid, 1863; Wakely, 1935; Doets, 1952; Beirne, 1954; Bretherton, 1956; Hannemann, 1977b; Buszko, 1986; Gibeaux & Picard, 1992), *I. bifrons* L. (Gibeaux & Picard, 1992), *I. montana* L. (Gibeaux, 1989d; Bigot et al., 1990; Gibeaux & Picard, 1992) *I. hirta* L. (Gibeaux & Picard, 1992) and *Buphthalmum salicifolium* L. (Hannemann, 1977b; Arenberger & Jaksic, 1991; Gibeaux & Picard, 1992). Another recorded hostplant is *Carlina vulgaris* L. (Beirne, 1954). The larva of the first generation lives in the stems, near the junction with a leaf. The second generation in the flower buds and on the seed-heads eating into the pith of the stem. Pupation in the excavated stem parts.

REMARKS.– The species shows considerable variation in size. It seems that specimens from cooler climates are larger than those from warmer areas.

The paratype of *L. inulaevorus* placed in my possession by Mr Gibeaux, shows genital structures identical with those of the present species.

100. Hellinsia chrysocomae (Ragonot, 1875)

Leioptilus chrysocomae Ragonot, 1875: 74.   Pl. 10: 12. Figs 114, 250.
Oidaematophorus bowesi Whalley, 1960: 29.

DIAGNOSIS.– Wingspan 18 mm. Colour bright yellow. From the wingbase to the base of the cleft a pale brown line. A dark spot at the base of the cleft.

VARIATION.– The line from the wing base may also extend toward the apex of the first lobe.

MALE GENITALIA.– Left valve lanceolate, with a slender saccular spine half as long as the valve. Right valve with a sacculus projecting beyond the valve and with a pointed apex.

FEMALE GENITALIA.– Ostium and antrum in left lateral side , narrow. Bursa copulatrix vesicular with some spiculation in the proximal half. Papillae anales well developed. Apophyses anteriores very short.

DISTRIBUTION.– England, France, Germany and Poland.

BIOLOGY.– The moth flies in August. The hostplants are *Aster linosyris* L. (Hannemann, 1976; Emmet, 1979; Gibeaux & Picard, 1992), *A. sedifolius* L. (Gibeaux & Picard, 1992) and *Solidago virgaurea* L. (Hannemann, 1976; Emmet, 1979; Bigot et al., 1990; Gibeaux & Picard, 1992;). The life-cycle undescribed.

101. Hellinsia osteodactylus (Zeller, 1841)

Pterophorus osteodactylus Zeller, 1841: 851.   Pl. 11: 1. Figs 115, 251.
Leioptilus cinerariae Millière, 1869: 418.

DIAGNOSIS.– Wingspan 16-23 mm. Colour yellow-white to bright yellow. At the base of the cleft a small spot and a costal dash beyond the base of the cleft in the first lobe.

MALE GENITALIA.– Left valve with a broadly rounded apex; the sacculus with a slender spine about one third as long as the valve. The sacculus of the right valve has the apex bilobate and club-like.

FEMALE GENITALIA.– Ostium laterally excavated into the 7th tergite. Antrum small, almost squarish. Bursa copulatrix and vesica seminalis vesicular, without a signum.

DISTRIBUTION.– Almost throughout Europe. Extends into Asia Minor and North Africa, and eastwards as far as Japan.

BIOLOGY.– The moth flies from May to September. The hostplants are *Solidago virgaurea* L. (Mitterberger, 1912; Beirne, 1954; Gozmány, 1962; Hannemann, 1977b; Buszko, 1986; Gibeaux & Picard, 1992), *Senecio nemorensis* L. (Hannemann, 1976), *S. fuchsii* C.C. Gmelin (Mitterberger, 1912; Hannemann, 1976; Buszko, 1986), *S. bicolor* Willd. (Emmet, 1979; Nel, 1989c; Gibeaux & Picard, 1992) and *Aster linosyris* Bernh. (Hannemann, 1976; Arenberger & Jaksic, 1991). The larva feeds on the flowers and the seeds. It hibernates among detritus and pupates in spring.

102. *Hellinsia pectodactylus* (Staudinger, 1859)

Pterophorus pectodactylus Staudinger, 1859: 258.　　　　Pl. 11: 2. Figs 116, 252.
Lioptilus angustus Walsingham, 1880: 43.
Lioptilus stramineus Walsingham, 1880: 41.
Pterophorus melanoschisma Walsingham, 1908: 920.

DIAGNOSIS.– Wingspan 16-20 mm. Colour bright yellow. At the base of the cleft a small brown spot.

MALE GENITALIA.– Left valve with a rather stout saccular spine one fourth as long as the valve. Sacculus of right valve with a club-like apex, slightly projecting beyond apex of the valve.

FEMALE GENITALIA.– Ostium and antrum hardly sclerotised, almost centrally placed. The 8th tergite rectangular, with short apophyses anteriores.

DISTRIBUTION.– South-West Germany, France, Spain, Canary Islands and Greece. In North America the species occurs in the southern half of the continent and the Rocky Mountains up to 2750 m.

BIOLOGY.– On the mainland the moth flies in May and June, later from August to October. The hostplants are *Solidago virgaurea* L. (Mitterberger, 1912; Gozmány, 1962; Hannemann, 1977b) and *Aster linosyris* Bernh. (Mitterberger, 1912; Gozmány, 1962; Hannemann, 1977b). On the Canary Islands it flies in March and April. The hostplant there is *Phagnalon saxatile* L. (Walsingham, 1908). The larva feeds on the flowers.

103. *Hellinsia distinctus* (Herrich-Schäffer, 1855)

Pterophorus distinctus Herrich-Schäffer, 1855: 379.　　　　Pl. 11: 3. Figs 117, 253.
Pterophorus sibericus Caradja, 1920: 86.
Leioptilus zermattensis Müller-Rutz, 1934: 118.

DIAGNOSIS.– Wingspan 15-20 mm. Colour from pale yellow-white to yellow-brown, speckled with red-brown scales. Before the base of the cleft a small spot; in the first lobe a costal spot obliquely beyond the spot near the cleft, and a smaller subapical spot at the costa. There are no distinct spots at the dorsum of the second lobe.

MALE GENITALIA.– Left valve slightly rounded, in some populations with the apex more acute. The saccular spine is slender and nearly half as long as the valve. The right valve is more acute, with a small saccular spine.

FEMALE GENITALIA.– Ostium laterally placed, with small sclerotised protrusions in the apex of the antrum. The bursa copulatrix vesicular, with numerous minute spiculae; the highest density is found in the equatorial plane of the bursa. The 8th tergite a little wider than long, with rather well developed apophyses anteriores.

DISTRIBUTION.– North, West, Central and South-East Europe, extending into Russia.

BIOLOGY.– The moth flies in July and August. The hostplants are *Gnaphalium sylvaticum* L. (Frey, 1886; Hannemann, 1977b; Buszko, 1986), *Artemisia absinthium* L. (Gozmány, 1962; Hannemann, 1977b) and *A. chamaemelifolia* Vill. (Gibeaux & Picard, 1992). These plants grow in shaded localities in woodland areas. In Scandinavia and Denmark in open land, often near the coast. The larva feeds until October on the flower and unripe seeds. Pupation occurs on the ground in a spinning, often several specimens together.

104. Hellinsia didactylites (Ström, 1783)

Phalaena Alucita didactylites Ström, 1783: 89.   Pl. 11: 4. Figs 118, 254.
Alucita scarodactyla Hübner, [1813]: t. 4, figs 21, 22.
Alucita icarodactyla Treitschke, 1833: 247.

DIAGNOSIS.– Wingspan 19-23 mm. Colour grey-white to grey-yellow; some specimens yellow. Before the base of the cleft a small brown spot; along the costa of the first lobe a brown dash, faintly interrupted at one third.

MALE GENITALIA.– Left valve rahter wide, with a very long and slender saccular spine, which projects beyond the valve. Right valve narrower, with a very weakly developed saccular spine.

FEMALE GENITALIA.– Ostium poorly developed. Antrum narrow, funnel-shaped, with two small, laterally placed, sclerotised plates. The 7th tergite longitudinally rectangular, with small apophyses anteriores.

DISTRIBUTION.– North, West, Central and East Europe, and Italy. Absent from England.

BIOLOGY.– The moth flies from May to August, in two generations. The hostplants are *Hieracium umbellatum* L. (Hannemann, 1977b), *H. prenanthoides* Villiers (Bigot et al., 1990; Gibeaux & Picard, 1992), *H. amplexicaule* L. (Gibeaux & Picard, 1992), *H. murorum* L. (Mitterberger, 1912; Hannemann, 1977b; Buszko, 1986), *H. lachenalii* Gmel. (Buszko, 1986) and *H. sylvaticum* L. (Hannemann, 1977b). These hostplants grow on sandy, dry places, usually on poor soils. The larva feeds on the developing flowers. The winter brood hibernates in a spinning at the plant and pupates in the late winter or early spring among detritus or leaves.

REMARKS.– Mr. O. Karsholt drew my attention to the description of the nominate species. He has shown that the description could only fit the species previously known as *H. scarodactyla* Hübner (Gielis, 1993).

105. *Hellinsia tephradactyla* (Hübner, [1813])

Alucita tephradactyla Hübner, [1813]: t. 4, fig. 17.   Pl. 11: 5. Figs 119, 225.

DIAGNOSIS.– Wingspan 18-23 mm. Colour grey-white, speckled with numerous black and brown scales. These scales are grouped in three or four longitudinal rows, extending from the wing base into the lobes. The fourth row of scales is often ill-developed dorsally in the second lobe. At the costa of the first lobe and at the dorsum of both lobes some small, hardly recognisable spots.

MALE GENITALIA.– Left valve with a slender curved saccular spine one third as long as the valve. Right valve less rounded and with a small saccular spine.

FEMALE GENITALIA.– Ostium oval. Antrum funnel-shaped, one and a half times as long as wide. In the vesicular bursa copulatrix a large double signum, each composed of well developed spines.

DISTRIBUTION.– North, West, Central and East Europe and Italy.

BIOLOGY.– The moth flies in June and July. The hostplants are *Solidago virgaurea* L. (Mitterberger, 1912; Beirne, 1954; Gozmány, 1962; Hannemann, 1977b; Emmet, 1979), *Aster bellidiastrum* Scop. (Mitterberger, 1912; Hannemann, 1977b; Emmet, 1979; Gibeaux & Picard, 1992) and *Bellis perennis* L. (Mitterberger, 1912; Bigot et al., 1990; Arenberger & Jaksic, 1991; Gibeaux & Picard, 1992). The hostplants grow on shady spots in open forests. The larva feeds by night on the leaves, forming holes. By day they rest along the stems. Pupation on the plant.

106. *Hellinsia lienigianus* (Zeller, 1852)

Pterophorus lienigianus Zeller, 1852: 380.   Pl. 11: 6; pl. 16: 8. Figs 120, 256.
Pterophorus melinodactylus Herrich-Schäffer, 1855: 71.
Pterophorus scarodactylus Becker, 1861: 56.
Leioptilus serindibanus Moore [in: Walsingham], 1887: 527.
Leioptilus sericeodactylus Pagenstecher, 1900: 240.
Ovendenia septodactyla Tutt (nec Treitschke), 1905a: 37.
Pterophorus victorianus Strand, 1913: 130.
Pterophorus hirosakianus Matsumura, 1931: 1056.

DIAGNOSIS.– Wingspan 17-21 mm. Forewing colour grey-white. Markings consist of an arched spot around the base of the cleft, at the costa of the first lobe two spots, the basal one the largest. Fringes grey-brown.

VARIATION.– There is some variety in the colour from grey-white to ferrugineous white, giving the species a more reddish appearance.

MALE GENITALIA.– The right valve with a minute spine in the middle of the valve along the saccular margin. The left valve with a short saccular process beginning in the centre of the valve, the bending towards the dorsal margin, and extending towards the tip of the valve.

FEMALE GENITALIA.– Antrum simple, placed in the left lateral side. Directly beneath the antrum, the ductus bursae is divided into a long and slender vesicular ductus seminalis and a gradually widening ductus bursae. The bursa copulatrix vesicular. Signum double, in shape of two groups of concentrically arranged minute spiculae.

DISTRIBUTION.– The entire area, uncommon around the Mediterranean sea. Also in the east Palaearctic region, North America, India, South-East Asia and Africa.

BIOLOGY.– The moth flies from June to August. The hostplants are *Artemisia vulgaris* L. (Doets, 1946; Gozmány, 1962; Hannemann, 1977b; Aarvik, 1987; Arenberger & Jaksic, 1991; Gibeaux & Picard, 1992; Gielis, bred) and *Leucanthemum vulgare* Lam. The larva feeds on the leaves, spinning the top and sides of the leaves together and making a shelter from which it eats the underside, causing window-like spots. Later the larva feeds on the upper leaves. Pupation on the underside of a leaf. The hostplants are found in the shade of trees, and sunny places are avoided (Beirne, 1954). Other recorded hostplants are: *Artemisia campestris* L. (Becker, 1861), *Tanacetum* sp. (Sutter, 1991) and possibly *Solanum* sp. (Sutter, 1991).

Oidaematophorus Wallengren, 1862

Oidaematophorus Wallengren, 1862: 19.
  Type species: *Alucita lithodactyla* Treitschke, 1833; monotypy.
Oedaematophorus Zeller, 1867; emendation.
Ovendenia Tutt, 1905: 37.
  Type species: *Alucita septodactyla* Treitschke, 1833; original designation.

DIAGNOSIS.– Forewing broad, in general well marked. The species are larger than those in the previous genus. The middle leg with distinct scale-brushes around the base of the spur pairs.

MALE GENITALIA.– Asymmetrical valvae, with spines and saccular processes.

FEMALE GENITALIA.– In this genus, the antrum is distinctly wide, inverted bell-shaped, and heavily sclerotised.

DISTRIBUTION.– Known from all zoographical regions. The majority of species seems to occur in the New World.

BIOLOGY.– The hostplants belong in general to the Asteraceae.

107. *Oidaematophorus lithodactyla* (Treitschke, 1833)

Alucita lithodactyla Treitschke, 1833: 245.   Pl. 11: 7; pl. 16: 4, 5. Figs 121, 257.
Alucita septodactyla Treitschke, 1833: 246.
Pterophorus similidactylus Dale, 1834: 263.
Pterophorus phaeodactylus Stephens, 1835: 375.
Pterophorus lithoxylodactylus Duponchel [in: Godart], 1840a: 670.

DIAGNOSIS.– Wingspan 26-29 mm. Colour grey-brown to reddish-brown. Before the base of the cleft a wedge-like spot extending to the disc; at the costa above the base of the cleft a longitudinal spot; in both lobes a longitudinal dark group of scales which form a spot.

MALE GENITALIA.– Left valve with a curved saccular spine, one third to one half as long as the valve. Right valve without spines or thorns. Both valves ending in a small acute tip.

FEMALE GENITALIA.– Ostium plate smoothly asymmetrically excavated. The antrum rounded. The distal part of the ductus bursae with a weakly sclerotised protrusion.

DISTRIBUTION.– Europe, except for the far south of Spain and Italy. Extending into Asia Minor and known from Japan.

BIOLOGY.– The moth flies in July and August. The hostplants are *Inula salicina* L. (Mitterberger, 1912; Hannemann, 1977b; Aarvik et al., 1986; Buszko, 1986), *I. conyza* L. (Beirne, 1954; Gozmány, 1962; Hannemann, 1977b; Nel, 1986a; Gibeaux & Picard, 1992; Gielis, bred), *I. germanica* L. (Hannemann, 1977b), *I. hirta* L. (Gibeaux & Picard, 1992), *I. helenioides* DC. (Gibeaux & Picard, 1992) and *Pulicaria dysenterica* Bernh. (de Graaf, 1859; South, 1881; Mitterberger, 1912; Beirne, 1954; Gozmány, 1962; Hannemann, 1977b; Gibeaux & Picard, 1992). The plants grow in fairly dry to wet meadows and open forest localities. The larva prefers the plants growing in shady places. It feeds on flowerbuds, the stem and the leaves. Holes appear in the leaves. Commonly several larvae are present on each plant. Pupation on the lower parts of the hostplant.

108. *Oidaematophorus rogenhoferi* (Mann, 1871)

Pterophorus rogenhoferi Mann, 1871: 79.   Pl. 11: 8. Figs 122, 258.

DIAGNOSIS.– Wingspan 30-32 mm. Colour brown-grey. A wedge-like, obliquely ending, terminal spot before the base of the cleft; a black costal spot above the base of the cleft; a small central longitudinal spot in the centre of the first lobe; a small fringe spot at three fourths of the first lobe; black scale-groups in the basal two thirds of the forewing and in the second lobe.

MALE GENITALIA.– Left valve wider than right valve. Both valves with rounded apices. The saccular spine in the left valve longer than in *O. lithodactyla*.

FEMALE GENITALIA.– Ostium positioned obliquely to the axis of the antrum. Antrum gradually narrowing. Vesica seminalis appears as a large protrusion of the junction area between the ductus bursae and the bursa copulatrix.

DISTRIBUTION.– Scandinavia and the Alps of Central Europe.

BIOLOGY.– The moth flies in July and August. Nel (1986a), in France, bred the species from *Erigeron acer angulosus* Gaudin; Burmann (1944), in Austria, from *E. alpinus* L.; and Karsholt, in Norway, from *E. acer politus* Fries (Aarvik et al., 1986).

REMARKS.– Differs from *O. lithodactyla* by its larger size and darker colour.

109. *Oidaematophorus constanti* (Ragonot, 1875)

Oedematophorus constanti Ragonot, 1875: 205.   Pl. 11: 9. Figs 123, 259.

DIAGNOSIS.– Wingspan 26-29 mm. Colour reddish-yellow, speckled with numerous black scales. Before the base of the cleft an oblique spot, terminally margined ochreous, beyond this spot at the costa a longitudinal spot. The wing shape more elongated than in *O. lithodactyla*.

MALE GENITALIA.– Left valve wider than right valve, both apices rounded. The long saccular spine is curved at the base and with another loop near the apex of the valve. Right valve with an indistinct saccular spine.

FEMALE GENITALIA.– Ostium with a central protrusion. Antrum wide, gradually narrowing.

DISTRIBUTION.– Mountainous areas of Central Europe.

Biology.– The moth flies in June and July. The hostplants are *Inula montana* L. (Mitterberger, 1912; Hannemann, 1977b; Nel, 1986a; Bigot et al., 1990; Gibeaux & Picard, 1992), *I. vaillantii* Vill. (Hannemann, 1977b), *I. hellenium* L. (Hannemann, 1977b), *I. helenioides* DC. (Gibeaux & Picard, 1992), *I. oculus christi* L. (Hannemann, 1977b; Marek & Skyva, 1985) and *I. conyza* DC. (Hannemann, 1977b; Arenberger & Jaksic, 1991; Gibeaux & Picard, 1992). The larva feeds on the leaves making holes in them.

REMARKS.– See *O. giganteus*.

110. *Oidaematophorus giganteus* (Mann, 1855)

Pterophorus giganteus Mann, 1855: 570.   Pl. 11: 10. Figs 124, 260.

DIAGNOSIS.– Wingspan 28-31 mm. Colour brown-yellow. A small spot before the base of the cleft, becoming fainter in its extension to the costa beyond the base of the cleft. Both lobes acute, and curved dorsally.

MALE GENITALIA.– Left valve elongated, with a simple, nearly straight spine which is half as long as the valve. Right valve of the same shape, but without a saccular spine.

FEMALE GENITALIA.– Ostium plate flattened, without a protrusion. The antrum gradually narrowing, as in *O. constanti*.

DISTRIBUTION.– Southern France, Italy and Corsica.

BIOLOGY.– The moth flies in June and July. The hostplants are *Inula hellenium* L. (Legrand, 1936; Gibeaux & Picard, 1992) and *Pulicaria odora* Reichenb. (Nel, 1986a; Gibeaux & Picard, 1992). The hostplants grow in scrubby localities in the warmer mediterranean area.

REMARKS.– Externally the species resembles *O. constanti*, but it is larger and has a curved acute tip of the forewing lobes. Also the saccular spine is differently arranged.

The smaller *O. vafradactylus* has a more curved saccular spine in the left valve, and the ostium plate is contrastingly deeply excavated.

111. *Oidaematophorus vafradactylus* Svensson, 1966

Oidaematophorus vafradactylus Svensson, 1966: 183.   Pl. 12: 1. Figs 125, 261.

DIAGNOSIS.– Wingspan 23-24 mm. Colour unicolourous brown. Along the costa speckled with blackish scales; before the base of the cleft an oblique spot and a spot at the costa above the base of the cleft.

MALE GENITALIA.– Left valve lanceolate with a curved saccular spine. The right valve simple.

FEMALE GENITALIA.– Ostium centrally positioned, deeply excavated in the antrum. The antrum one and a half times as long as wide, gradually narrowing.

DISTRIBUTION.– Sweden (Gotland and Öland).

BIOLOGY.– The moth flies in July and the beginning of August. The presumed hostplant is *Inula salicina* L.

REMARKS.– See *O. giganteus*.

Emmelina Tutt, 1905

Emmelina Tutt, 1905: 37.
   Type species: *Phalaena Alucita monodactyla* Linneaus, 1758; original designation.

DIAGNOSIS.– Palpi slender, erect. Second and 3rd abdominal segments elongated. The medial of the proximal pair of spurs on the hind legs is twice as long as the lateral spur. Forewing R2 and R3 fused, M3 and Cu1 stalked.

MALE GENITALIA.– Extremely asymmetrical, with numerous complex processes from the sacculus, cucullus, or originating in the medial part of the valvae.

FEMALE GENITALIA.– Ostium and antrum complex, located on the margin of the 7th tergite. Ductus bursae and ductus seminalis separate. Bursa copulatrix without a signum.

DISTRIBUTION.– Representatives are known from all regions of the world.

BIOLOGY.– The species seem to be polyphagous. However, preference is given to representatives of the genera *Convolvulus* and *Calystegia*.

112. *Emmelina monodactyla* (Linnaeus, 1758)

Phalaena Alucita monodactyla Linnaeus, 1758: 542.   Pl. 12: 2, 3. Figs 126, 262.
Phalaena bidactyla Hochenwarth, 1785: 336.
Pterophorus cineridactylus Fitch, 1854: 848.
Pterophorus naevosidactylus Fitch, 1854: 849.
Pterophorus pergracilidactylus Packard, 1873: 266.
Pterophorus barberi Dyar, 1903: 228.
Pterophorus pictipennis Grinnell, 1908: 320.

DIAGNOSIS.– Wingspan 18-27 mm. Colour pale ferrugineous. Markings dark brown, consisting of a small longitudinal spot near the base of the apex. Centrally in the costa of the first lobe a small dark spot and two distinct subapical and apical spots at the dorsal margin of the first lobe. In the outer margin of the second lobe three spots are present, near the anal angle, the apex and almost centrally placed.

VARIATION.– The species is highly variable in its colour, from a pale grey-white with faint markings to a dark ferrugineous brown. Also the intensity of the markings may vary, and there is a considerable variation in size.

MALE GENITALIA.– Valvae asymmetrical. The left valve is rounded and rather wide, and has a complex structure of saccular processes, forming three elements. Two of these consist of a slender arm, orginating at one third of the valve. The longer reaches towards the apex of the valve and the shorter has only half of its length. The third segment originates rather narrow, then gradually widens, before branching into a short slender arm and a longer arm curved in a "S"-shape. The right valve is rather lanceolate, pointed towards the apex. At two thirds of the valve-length a slender saccular arm is present, slightly shorter than the apex of the valve.

FEMALE GENITALIA.– Antrum acentrically placed with a simple ostium. Next to the ostium is the opening of the ductus seminalis, giving the distal margin of the 7th sternite an asymmetrical bulged appearance. The ductus bursae is straight and slender, progressing into the vesicular bursa copulatrix, which is without a signum.

DISTRIBUTION.– Europe, Africa, Asia, North America and the northern half of Central America.

BIOLOGY.– The moth flies throughout the year. Reported hostplants are *Convolvulus arvensis* L. (Amsel, 1935b; Gozmány, 1962; Hannemann, 1977b), *Calystegia soldanella* (L.) R. Br. (Nel, 1989c), *C. sepium* (L.) R. Br. (Hannemann, 1977b; Parrella & Kok, 1978; Emmet, 1979), *Chenopodium* sp. (Hannemann, 1977b), *Atriplex* sp. and *Ipomoea batatas* L. (Yano, 1963; Parrella & Kok, 1978).

113. *Emmelina argoteles* (Meyrick, 1922)

Pterophorus argoteles Meyrick, 1922: 549.   Pl. 12: 4. Figs 127, 263.
Pterophorus jezonicus Matsumura, 1931: 1057.
Pterophorus komabensis Matsumura, 1931: 1057.
Pterophorus menoko Matsumura, 1931: 1057.
Pterophorus yanagawanus Matsumura, 1931: 1058.
Emmelina jezonica pseudojezonica Derra, 1987: 71.

DIAGNOSIS.– Wingspan 18-23 mm. Colour and markings as in *E. monodactyla*. Cannot be identified on external characters.

MALE GENITALIA.– The left valve has a large, broadened cucullar process dominating the apical part of the valve; the S-shaped saccular process is less developed than in *E. monodactyla*. The right valve has a large rounded apex, in contrast to the acute apex in *E. monodactyla*, and the saccular process is smaller.

FEMALE GENITALIA.– The ostium and antrum structures are wider and the ostium is more developed. The differences from *E. monodactyla* are small.

DISTRIBUTION.– Germany, France, Austria, Hungary, Japan and the nominate form from China.

BIOLOGY.– The moth flies in April, June, August and September. The hostplants are *Calystegia sepium* (L.) R. Br. (Nel & Prola, 1991), and *Ipomoea batatas* L. (Yano, 1963). The larva seems to feed on the flowers, as in *E. monodactyla.*

REMARKS.– There is some difference in the genital structure between specimens from the eastern Palaearctic region and those from Europe. Considering the differences Aren-berger (1981) demonstrated in the male genitalia of *Pterophorus* spp. if comparing West Palaearctic with Central Palaearctic specimens, I regard *E. pseudojezonica* as a synonym of *E. argoteles,* and to treat it as a subspecies is a matter of taste, which I personally oppose to.

Adaina Tutt, 1905

Adaina Tutt, 1905: 37.
  Type species: *Alucita microdactyla* Hübner, [1813]: pl. 5, figs 26, 27; original designation.

DIAGNOSIS.– Forewing without anal angles of the lobes. Wing venation: R1 separate; R2+3 and R4 stalked, with the base at the apical angle of the discal cell; R5 separate.

MALE GENITALIA.– Valvae asymmetrical. At the sacculus of the left valve a well developed spine or thorn, at the sacculus of the right valve a small thorn or ridge.

FEMALE GENITALIA.– Antrum laterally positioned. Ductus seminalis well developed, bursa copulatrix vesicular.

DISTRIBUTION.– *Adaina microdactyla* occurs throughout the Palaearctic region and ex-tends into the Indo-Australian fauna. In Africa four additional species are recognized and in the New World further 21 species.

BIOLOGY.– See *A. microdactyla.*

REMARKS.– The genus is closely related to the *Oidaematophorus, Pselnophorus* and *Pterophorus* group of genera, differing in the wing venation.

114. Adaina microdactyla (Hübner, [1813])

Alucita microdactyla Hübner, [1813]: pl. 5, figs 26, 27.   Pl. 12: 5. Figs 128, 264.
Adaina montivola Meyrick [*in*: Caradja & Meyrick], 1937: 170.

DIAGNOSIS.– Wingspan 13-17 mm. Colour pale yellow to yellow-white. Markings brown: a small spot at the base of the cleft; two spots at the costa of the first lobe at one fifth and two thirds; some dark scales at the dorsum of the first lobe; three small scale groups at the dorsum of the second lobe, the last near the apex of this lobe.

MALE GENITALIA.– Left valve rounded, with a distinct curved saccular spine. Right valve more elongated with two small spines along the saccular margin at one third and one half of the valve length.

FEMALE GENITALIA.– Ostium flattened, laterally placed in the 8th tergite. Antrum gradually narrowing, with two small lateral sclerotised plates. Ductus seminalis characteristically spiral-shaped.

DISTRIBUTION.– Europe, except for the far north of Scandinavia. Extending to the east into Asia Minor, Iran, and known from Japan, Solomon Islands and Indonesia.

BIOLOGY.– The moth flies from April to June and again from July to September. The hostplant is *Eupatorium cannabinum* L. (Beirne, 1954; Mellini, 1954; Gozmány, 1962; Hannemann, 1977b; Buszko, 1986; Gielis, bred). The plant grows along brooks, ponds, lakes and rivers, and is also found in marshes. The larva of the winter brood bores into the stem of the plant and feeds on the pith. This causes the formation of characteristic galls, often several in one stem. The larva hibernates in the gall and the moth emerges in spring. The summer brood is thought to feed on the flowers and seeds of the plant. These larvae pupate on the stem, or amongst debris on the soil surrounding the hostplant.

Calyciphora Kasy, 1960

Calyciphora Kasy, 1960: 175.
Type species: *Alucita xanthodactyla* Treitschke, 1833; original designation.

DIAGNOSIS.– The forewings are well marked; a dard brush present in the fringes of the second lobe near the anal angle. The venation shows the presence of R2 and a stalked R4 + R5. Sternite 8 of female forms a shield-like plate. Along the distal margin of this plate are numerous well developed scales present.

MALE GENITALIA.– The valvae are asymmetrical and usually possess cucullar protrusions. The saccular processes in the left valve is curved and long. The right valve is without a saccular process or this is very small. The aedeagus may have a curled tip.

FEMALE GENITALIA.– The 8th tergite is heavily sclerotised and in shape of a centrally indentated shield, which is laterally flattened. Ductus seminalis arises from the bursa copulatrix.

DISTRIBUTION.– Palaearctic region, but absent from North and North-West Europe.

BIOLOGY.– The hostplants belong to the Asteraceae, especially species of thistles.

115. *Calyciphora punctinervis* (Constant, 1885)

Aciptilia punctinervis Constant, 1885: 14.   Pl. 12: 6. Figs 129, 265.
Alucita tyrrhenica Amsel, 1954b: 6.

DIAGNOSIS.– Wingspan 16-19 mm. Colour pale yellow-white. At the dorsum of the second lobe two small black spots.

MALE GENITALIA.– Left valve wide, rounded, with a saccular process one third as long as the valve. Right valve narrower than left valve, with a small hook-like process.

FEMALE GENITALIA.– Ostium narrow. Antrum slender and progressing into the hardly sclerotised ductus bursae and bursa copulatrix. No signum. Apophyses anteriores absent. Apophyses posteriores three times as long as the papillae anales.

DISTRIBUTION.– Southern France, Corsica, Sardinia, Spain, Portugal.

BIOLOGY.– The moth flies in August and September. The hostplant is *Carlina corymbosa* L. (Nel, 1989c; Bigot et al., 1990).

REMARKS.– The established synonymy of *C. tyrrhenica* is based on the description and the illustration of the male genitalia by Amsel (1954).

116. *Calyciphora homoiodactyla* (Kasy, 1960)

Aciptilia (Calyciphora) homoiodactyla Kasy, 1960: 185.　　　　Pl. 12: 7. Figs 130, 266.

DIAGNOSIS.– Wingspan 18-23 mm. Colour yellow-white, along the costa and near the wing base some faint browning, the apex of both lobes and the fringes at the dorsal part of the apex darkened. Fringes at the centre of the dorsum of the second lobe with a dark brush.

MALE GENITALIA.– Left valve rounded, with strongly curved saccular process. Right valve elongated, the apex flattened, and with a small saccular spine in the centre of the valve.

FEMALE GENITALIA.– Ostium funnel-shaped, flattened. Ductus bursae slender. The antrum in a circular lamina antevaginalis.

DISTRIBUTION.– France, ex-Yugoslavia, Greece, and further into Asia Minor.

BIOLOGY.– The moth flies in July and August. In Greece the hostplant is an *Echinops* sp. (Wasserthal, 1970; Arenberger, 1983, without indication on the life-cycle). In the French Alps, Bigot and Picard (1983a) mention *Echinops ritro* L., which grows on the mountain slopes, as the hostplant.

117. *Calyciphora adamas* (Constant, 1895)

Aciptilia adamas Constant, 1895: 54　　　　Pl. 12: 8. Figs 131, 267.

DIAGNOSIS.– Wingspan 20-22 mm. Colour blue-grey. Markings consist of black scales: at the costa at one fourth, along the dorsum of the first lobe, and the costa of the second lobe, from the base of the cleft to the disc. Fringes grey-white, with a dark dash in the middle of the dorsum of the second lobe. The grey-brown abdomen has a dark brown dorsal longitudinal line.

MALE GENITALIA.– Left valve gradually narrowing, with a waved shape of the cucullar margin; a long cucullar process present; a saccular process centrally in the valve has a sharp bend towards tip. Right valve gradually widening, with a flat apex. Aedeagus long, curved.

FEMALE GENITALIA.– Ostium excavated. Antrum funnel-like, progressing into the long, slender ductus bursae. The proximal margin of the 8th tergite with large and wide apophyses anteriores, embracing the lamina antevaginalis and the 7th tergite.

DISTRIBUTION.– Spain, southern France, Italy.

BIOLOGY.– The moth flies in May and June, later again in September and the beginning of October. The hostplant is *Staehelina dubia* L. (Nel, 1988b; Bigot et al., 1990).

118. *Calyciphora acarnella* (Walsingham, 1898)

Alucita acarnella Walsingham, 1898: 131.   Pl. 12: 9. Figs 132, 268.

DIAGNOSIS.– Wingspan 21-24 mm. Colour pale brown-grey. A brown spot at the costa and dorsum of the wing at the base of the cleft. Fringes near the apex of both lobes brown-tipped.

MALE GENITALIA.– Left valve elongate and rounded, with a cucullar process at three fourths; a long double-looped saccular process extending beyond the apex of the valve. The right valve narrower, elongated, with a cucullar process at three fourths, and a sclerotised ridge from the base of the valve to the apex of the cucullar process.

FEMALE GENITALIA.– Ostium surrounded by a corona of thorn-like spiculae. 7th tergite with a flattened top and proximally terminating into a large bifurcate plate.

DISTRIBUTION.– Corsica, Sardinia.

BIOLOGY.– The moth flies from April to the beginning of August, probably in two generations. The hostplant recorded by Walsingham (1898) is *Picnomon acarna* (L.) Cass. (Nel, 1989c) and Nel (1989c), mentions *Ptilostemon casabonae* Greuter, growing in the mountains of the islands. Walsingham found the larvae on May 26th, moths emerged June 18-28th. Nel bred adults between April 27th and May 25th.

119. *Calyciphora albodactylus* (Fabricius, 1794)

Pterophorus albodactylus Fabricius, 1794: 348.   Pl. 13: 1. Figs 133, 269.
Pterophorus xerodactylus Zeller, 1841: 860.
Pterophorus xanthodactylus auct., (*nec* Treitschke, 1833).
Aciptilia sicula Fuchs, 1901: 72.

DIAGNOSIS.– Wingspan 20-28 mm. Colour pale straw-yellow. Before the base of the cleft a small brown spot; an indistinct discal spot and a longitudinal spot at one third of the costa of the first lobe. Fringes grey-white, with a distinctly brown hair-brush at one third of the dorsum of the second lobe.

MALE GENITALIA.– Left valve elongate, gradually narrowing. At the dorsal margin a hump at two thirds of the length, and a saccular process sharply curved and extending beyond the apex of the valve. Right valve with almost parallel margins, and a small cucullar process just before the tip.

FEMALE GENITALIA.– Ostium in the curved margin of the 7th tergite, which is club-shaped. Antrum funnel-shaped, rapidly tapering into the long and slender ductus bursae.

DISTRIBUTION.– Sweden, Finland, Central and South-West Europe, the Mediterranean area.

BIOLOGY.– The moth flies from June to September in the warmer parts of the area. In the northern areas in July and August. The hostplant is *Cirsium helenioides* L. (Kyrki & Karvonen, 1985; Nel, 1988b). Other reported hostplants are *Jurinea cyanoides* L. (Mitterberger, 1912; Hannemann, 1977b), *Cirsium ferox* (L.) DC. (Nel, 1988b), *Carlina vulgaris* L. (Fuchs, 1901; Mitterberger, 1912; Gozmány, 1962; Hannemann, 1977b; Nel, 1988b), *C. acanthifolia* All. (Nel, 1988b), *C. longifolia* Rchb. (Buszko, 1986) and

Echinops ritro L. (Bigot et al., 1990; Nel, 1988b). The hostplant localities are typically xerothermic places. The larva feeds on the underside of the leaves creating character-istic blotches, leaving the epidermis intact. The felt-hairs from the leaves are rolled up in small walls as in *P. galactodactylus*. Pupation under the leaves.

REMARKS.– The synonymy of *albodactylus* and *xerodactylus* was discussed by Karsholt & Gielis (1995).

120. Calyciphora xanthodactyla (Treitschke, 1833)

Alucita xanthodactyla Treitschke, 1833: 251.　　　　　　Pl. 13: 2. Figs 134, 270.
Aciptilia (Calyciphora) klimeschi Kasy, 1960: 177.

DIAGNOSIS.– Wingspan 20-27 mm. Colour white-yellow. Markings as in *C. albodactylus*.

MALE GENITALIA.– Left valve gradually narrowing; near the apex at the dorsal margin a stout, short cucullar process; the strongly backwardly curved saccular process slightly shorter than the valve. Right valve narrow, with almost parallel margins and a more slender process at the dorsal margin near the apex.

FEMALE GENITALIA.– Very like *C. albodactylus*, but differing in the wider ostium and the more spade-like shape of the 7th tergite.

DISTRIBUTION.– Poland and the Balkan countries.

BIOLOGY.– The moth flies in July and August. The hostplant is *Jurinea mollis* (L.) Reichenb. *Carlina longifolia* Rehb. (Buszko, 1974) is also stated; on this plant the larva feeds on the underside of the leaves, demonstrating a fenestrating feeding pat-tern.

121. Calyciphora nephelodactyla (Eversmann, 1844)

Alucita nephelodactyla Eversmann, 1844: 609.　　　　　　Pl. 13: 3. Figs 135, 271.
Aciptilia apollina Millière, 1883: 177.

DIAGNOSIS.– Wingspan 28-33 mm. Colour grey-white, speckled grey-brown. Markings as in *C. albodactylus*.

MALE GENITALIA.– Left valve as in *C. albodactylus*, but the dorsal process is medially placed and not at two thirds. The right valve has a longer dorsal process than in *C. albodactylus*, and is deeper cut out in the margin.

FEMALE GENITALIA.– As in *C. albodactylus*.

DISTRIBUTION.– Spain (Pyrenees), France, Italy, Austria, Poland, Czechoslovakia, Balkan countries, and extending into Russia.

BIOLOGY.– The moth flies in July and August, occurring on alpine meadows and rocky slopes from 1000-2000 m. The hostplant is *Cirsium eriophorum* L. (Zukowski, 1960; Jäckh, 1961; Burmann, 1965; Arenberger, 1983; Buszko, 1986; Nel, 1988b; Gielis, bred). The larva feeds on the underside of the leaves, causing a fenestrated pattern of frass. Pupation on the stem or near a vein under a leaf.

Porrittia Tutt, 1905

Porrittia Tutt, 1905: 37.
  Type species: *Alucita galactodactyla* [Denis & Schiffermüller], 1775; original designation.

DIAGNOSIS.– The forewing is well marked, with numerous spots and a dark fringe brush at the anal region of the dorsum of the second lobe. The venation shows R2 and R4 separate and a small M1.

MALE GENITALIA.– The left valve rounded, with a large curved saccular process. Right valve more elongate, with a saccular spine.

FEMALE GENITALIA.– Ostium almost flat, wider than the antrum. Antrum as long as the width of the ostium, progressing into the slender ductus bursae. Bursa copulatrix without a signum. Ductus seminalis at the junction area between the bursa copulatrix and the ductus bursae.

DISTRIBUTION.– Western and Central Europe.

BIOLOGY.– The hostplant is *Arctium*.

122. *Porrittia galactodactyla* ([Denis & Schiffermüller], 1775)

Alucita galactodactyla [Denis & Schiffermüller], 1775: 320.   Pl. 13: 4; pl. 16: 9. Figs 5, 136, 272.

DIAGNOSIS.– Wingspan 20-25 mm. Colour white. Markings black, and composed of a discal spot, a horizontal V-shaped spot at the base of the cleft, as an extension of the arms of the V a costal and dorsal spot, a small spot central at the costa of the first lobe, and a subterminal spot at the dorsum of both forewing lobes. Fringes white, with a black dash at the dorsum, near the base of the cleft. The hindwing lobes black-tipped.

MALE GENITALIA.– See genus description.

FEMALE GENITALIA.– See genus description.

DISTRIBUTION.– Western and Central Europe.

BIOLOGY.– The moth flies in June, rarely until August. The hostplant is *Arctium lappa* L. (Murray, 1957; Gozmány, 1962; Morris, 1963; Hannemann, 1977b; Buszko, 1986; Gielis, bred). The plant grows in waste places. The larva feeds on the underside of the leaves. It removes the felt-hairs of the plants creating ridges of these felt-hairs around windows and later holes made in the leaves. Pupation along the veins of the leaves or along the stem. Pupal period lasts 10-14 days.

REMARKS.– *Arctium nemorosum* Lej. is also mentioned as a hostplant (Buszko, 1986). In the Netherlands this plants is rejected by the larvae. The range of the moth is restricted to the limited distribution of *A. lappa*.

Merrifieldia Tutt, 1905

Merrifieldia Tutt, 1905: 37.
  Type species: *Phalaena Alucita tridactyla* Linnaeus, 1758; original designation.

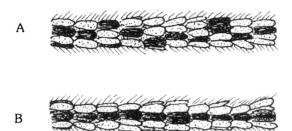

Fig. 14. Distribution of pale and dark scales on the antenna of A, *Merrifieldia leucodactyla* and B, *M. tridactyla*. (After Svensson).

DIAGNOSIS.– The forewing shows a dark line or shading at the costa. Some species have dark costal spots. The venation is reduced, a single radial vein (R4) is present, and in some species also a reduced R2 may be recognized.

MALE GENITALIA.– The valvae are asymmetrically shaped, and also the saccular processes are asymmetrical, being well developed in both valvae.

FEMALE GENITALIA.– Ostium simple. Antrum with a small sclerotised region distad of the ostium. Ductus bursae slender and long. Bursa copulatrix vesicular, with a pair of signa which may be strongly developed. Ductus seminalis originating in the centre of the ductus bursae.

DISTRIBUTION.– Except for the northern area, present in the entire Palaearctic region.

BIOLOGY.– The hostplants are Lamiaceae.

123. *Merrifieldia leucodactyla* ([Denis & Schiffermüller], 1775)

Alucita leucodactyla [Denis & Schiffermüller], 1775: 146. Pl. 13: 5. Figs 14A, 137, 273.
Alucita leucodactyla Hübner, [1805]: t. 1, fig. 2.
Alucita theiodactyla Hübner, [1825]: 431.
Aciptilia wernickei Wocke [in: Wernicke], 1898: 376.
Alucita fitzi Rebel, 1912: 107.
Alucita dryogramma Meyrick, 1930: 567.
Pterophorus tridactylus auct. (*nec* Linnaeus, 1758).
Pterophorus tetradactyla auct. (*nec* Linnaeus, 1758).

DIAGNOSIS.– Wingspan 18-25 mm. Colour straw-yellow, with a brown suffusion on the basal half of the wing, and at the costa in the centre of the first lobe. The main character to separate this species from the following is the presence of scales on the antennae: a single row of white scales margined by some rows of brown scales.

MALE GENITALIA.– Left valve rounded with basal section of saccular process wide and followed by a curved slender spine of the same length. Right valve more elongate, the wide basal part of the saccular process very long, progressing into an acutely hooked plough-like process reaching the cucullar margin.

FEMALE GENITALIA.– Ostium funnel-shaped, progressing into the narrow and slender antrum and ductus bursae. Bursa copulatrix with a pair of longitudinal, almost thread-like signa. The margin of the 7th tergite with a small and a long stout apophysis-like process.

DISTRIBUTION.– Except for the far north, in all of the area. North Africa, Asia Minor.

BIOLOGY.– The moth flies from May to September. The hostplants are *Thymus serpyllum* L. (Hannemann, 1977b; Buszko, 1986; Bigot et al., 1990; Arenberger & Jaksic, 1991), *T. pulegioides* L. (Buszko, 1986), *T. drucei* Ronniger (Emmet, 1979), *Pulmonaria officinalis* L. (Mitterberger, 1912; Hannemann, 1977b) and *Origanum vulgare* L. (Hannemann, 1977b). These plants grow on shrubby localities, downland, dunes and in light forests. The larva feeds from autumn, overwinters and feeds again in spring. The whole life cycle is spent on the plant.

124. *Merrifieldia tridactyla* (Linnaeus, 1758)

Phalaena Alucita tridactyla Linnaeus, 1758: 542.   Pl. 13: 6. Figs 14B, 138, 274.
Pterophorus fuscolimbatus Duponchel [in: Godart], 1844: 498.
Pterophorus icterodactylus Mann, 1855: 571.
Alucita icterodactyla noctis Caradja, 1920: 83.
Alucita baliodactyla menthae Chrétien, 1925: 243.
Alucita icterodactyla phillipsi Huggins, 1955: 124.
Aciptilia exilidactyla Buszko, 1975: 141.
Merrifieldia neli Bigot & Picard, 1989: [31].

DIAGNOSIS.– Wingspan 18-23 mm. Colour grey-white to bright yellow, suffused by numerous pale brown scales, especially in the basal half of the wing, and a narrow pale brown costal line. Antennae ringed white and dark brown.

MALE GENITALIA.– Left valve with a broad, hooked saccular process. The right valve with a more broadly built saccular process, which is curved towards the valve margin.

FEMALE GENITALIA.– Very similar to the genitalia of *M. leucodactyla*, but differing in the absence of the processes on the proximal margin of the 7th tergite.

VARIATION.– There is a considerable variation in the length and shape of the saccular processes in both valvae (Arenberger, 1981).

DISTRIBUTION.– Europe, except for the northern half of Scandinavia. Extending into Asia Minor and North Africa.

BIOLOGY.– The moth flies from April to September. In the southern parts of the area in two to four generations. The hostplants are *Thymus serpyllum* L. (Buszko, 1986; Heckford, 1988), *T. vulgaris* L. (Bigot & Picard, 1989; Nel, 1989d), *T. drucei* Ronniger, *T. pulegioides* L. (Heckford, 1988), *T. praecox* Opiz. (Heckford, 1988), *T. marschallianus* Willd. (Buszko, 1986; Bigot et al., 1990; Arenberger & Jaksic, 1991), *Mentha sylvestris* (= *longifolia* (L.) Hudson (Nel, 1991), *M. rotundifolia* (L.) Hudson (Nel, 1991) and *Calamintha nepeta* (L.) Savi (Nel, 1989d). The larva feeds on the flowers and flower-buds, later on the unripe seeds. The pupal period lasts approximately 10 days.

125. *Merrifieldia baliodactylus* (Zeller, 1841)

Pterophorus baliodactylus Zeller, 1841: 861.　　　　　Pl. 13: 7; pl. 16: 6, 7. Figs 139, 275.
Aciptilia baliodactyla var. *meridionalis* Staudinger, 1880: 432.

DIAGNOSIS.– Wingspan 20-27 mm. Colour yellowish white. Markings dark brown: an indistinct costal line from the wing-base to a distinct spot at the costa before the base of the cleft; also some ill-defined suffusion near the costa of the first lobe. Abdomen with three narrow pale brown longitudinal lines.

MALE GENITALIA.– Left valve rounded, with a rather short, stout, almost dorsally positioned saccular spine. The right valve with more parallel margins, and a saccular process, one third of valve length, with an angulated apex.

FEMALE GENITALIA.– Ostium narrow, slightly excavated. Antrum narrow, slender, hardly sclerotised, rapidly tapering into the long and slender ductus bursae. In the vesicular bursa copulatrix a pair of wedge-shaped signa, with the margins on the broad side rounded.

DISTRIBUTION.– Western, Central and South Europe. In the southern parts of the range more in mountainous areas, in the northern parts in lowland localities.

BIOLOGY.– The moth flies from June to August. The host plant is *Origanum vulgare* L. (Mitterberger, 1912; Hannemann, 1977b; Emmet, 1979; Nel, 1989d; Arenberger & Jaksic, 1991; Gielis, bred). In autumn the larva feeds into the central stem and root. In spring it feeds on the stem and leaves. Pupation on a leaf or along a stem.

126. *Merrifieldia malacodactylus* (Zeller, 1847)

Pterophorus malacodactylus Zeller, 1847: 905.　　　　　Pl. 13: 8. Figs 140, 276.
Pterophorus meristodactylus Zeller, 1852: 396.
Alucita indocta Meyrick, 1913a: 111.
Alucita subtilis Caradja, 1920: 81.
Alucita parca Meyrick, 1921: 421.
Alucita subcretosa Meyrick, 1922: 549.
Alucita phaeoschista Meyrick, 1923: 277.
Alucita spicidactyla Chrétien, 1923: 231.
Alucita rayatella Amsel, 1959: 55.
Aciptilia spicidactyla spp. *insularis* Bigot, 1961: 7.
Aciptilia livadiensis Zagulajev & Filippova, 1976: 40.
Pterophorus malacodactylus ssp. *transdanubinus* Fazekas, 1986: 13.
Merrifieldia garrigae Bigot & Picard, 1989: [33].
Merrifieldia moulignieri Nel, 1991: 174.
Merrifieldia inopinata Bigot, Nel & Picard, 1993: 122. **Syn. n.**

DIAGNOSIS.– Wingspan 15-27 mm. Colour yellow-white to straw-yellow. Markings consist of a brown costal line, widened in a spot before the base of the cleft. Specimens of *M. malacodactylus* are generally paler coloured and smaller than those of *M. baliodactylus*. However, very large pale specimens do exsist. The markings are less well developed, and the costal forewing spot is often integrated in the costal line.

MALE GENITALIA.– As in *M. baliodactylus*.

FEMALE GENITALIA.– As in *M. baliodactylus*.

DISTRIBUTION.– In the southern parts of the area and the Mediterranean region.

BIOLOGY.– The moth flies from April to October, probably in three generations. The hostplants are *Origanum vulgare* L., *Thymus herba-barona* Loisel (Nel, 1991), *Lavandula stoechas* L. (Nel, 1989d), *L. angustifolia* Miller (Nel, 1989d), *L. latifolia* Medicus (Bigot & Picard, 1989; Nel, 1989d), *Calamintha nepeta* (L.) Savi (Bigot & Picard, 1989; Nel, 1989d), *Rosmarinus officinalis* L. (Bigot & Picard, 1989; Nel, 1989d) and *Nepeta nepetellae* L. (Bigot et al., 1990; Arenberger & Jaksic, 1991).

REMARKS.– This species is difficult to separate from *M. baliodactylus*. The previous species tends to occur in the cooler parts of the area and in mountainous localities in the south. The present species mainly occurs in the Mediterranean area and in lowlands and valleys. It is possible that *M. baliodactylus* and *M. malacodactylus* should be regarded as conspecific, with two phenotypes caused by climatological and temperature conditions.

The species described by Bigot, Nel and Picard show no significant differences from *malacodactylus* and should be considered as synonyms.

127. *Merrifieldia semiodactylus* (Mann, 1855)

Pterophorus semiodactylus Mann, 1855: 570.   Pl. 13: 9. Figs 141, 277.

DIAGNOSIS.– Wingspan 17-20 mm. Colour greenish yellow. Three small yellow spots along the costa. Fringes at the dorsum of the second lobe chequered grey and black.

MALE GENITALIA.– Left valve more rounded than right valve. In the left valve a short, stout spine-like saccular process. This process has a basal curved spine pointing toward the base of the valve. Right valve with a small saccular thorn.

FEMALE GENITALIA.– Ostium wide, funnel-shaped and tapering into the gradually narrowing antrum. Bursa copulatrix vesicular, with a double, longitudinally spiculated signum. Apophyses anteriores absent. Apophyses posteriores three times as long as the papillae anales.

DISTRIBUTION.– Corsica, Sardinia. The records from southern France and Spain need confirmation.

BIOLOGY.– The moth flies in July. The hostplant is *Mentha suaveolens insularis* Greuter (Nel, 1989d).

128. *Merrifieldia hedemanni* (Rebel, 1896)

Gypsochares hedemanni Rebel, 1896: 115.   Pl. 13: 10. Figs 142, 278.
Alucita hesperidella Walsingham, 1908: 917.

DIAGNOSIS.– Wingspan 16-18 mm. Colour brown-yellow. Small costal spots at the base of the cleft, and centrally at the first lobe. At the dorsum of the second lobe a row of dark scales.

MALE GENITALIA.– Left valve rounded, with a short stout cucullar process. The right valve more elongate, with almost parallel margins, containing an angulated saccular process, having a rather broad apex.

FEMALE GENITALIA.– Ostium flat, hardly sclerotised. Antrum three times as long as wide, and tapering into the hardly sclerotised ductus bursae and bursa copulatrix. No signum. Apophyses anteriores absent. Apophyses posteriores four times as long as the papillae anales.

DISTRIBUTION.– Canary Islands.

BIOLOGY.– The moth flies from March to May. The hostplant is *Micromeria varia* Bentham; the larva feeds on the leaves.

REMARKS.– This species resembles *Puerphorus olbiadactylus* Millière very much, differing in the colour of the second lobe of the forewing, and the shape of the first lobe (here more acute).

129. *Merrifieldia chordodactylus* (Staudinger, 1859)

Pterophorus chordodactylus Staudinger, 1859: 259.   Pl. 14: 1. Figs 143, 279.
Pterophorus probolias Meyrick, 1891: 12.
Alucita particiliata Walsingham, 1908: 916.

DIAGNOSIS.– Wingspan 18-22 mm. Colour grey-yellow. Markings consist of a brown longitudinal line from the base to the apex of the second lobe and a line in the first lobe. Fringes grey-white, with a characteristic black edge, surrounding the wing as a black margin.

MALE GENITALIA.– Left valve rounded, with a saccular process half as long as the valve. This process is slightly curved, and widened at middle. The right valve less rounded, with a well developed wide and curved saccular process, ending in plough-like shape.

FEMALE GENITALIA.– Ostium narrow, excavated. Antrum long and slender, tapering into the slender ductus bursae. Bursa copulatrix vesicular, with a pair of signa composed of an accumulation of small sclerotised ridges.

DISTRIBUTION.– Canary Islands, Morocco, Algeria, southern Spain.

BIOLOGY.– The moth flies December to February, and in April and July. The hostplant is unknown, but Walsingham (1908) mentions *Lavandula abrotanoides* Lamarck as a possible hostplant, and Chrétien (1923) mentions *L. multifida* L.

130. *Merrifieldia bystropogonis* (Walsingham, 1908)

Alucita bystropogonis Walsingham, 1908: 915.   Pl. 14: 2. Figs 144, 280.

DIAGNOSIS.– Wingspan 16-20 mm. Colour pale brown-grey, the apical half of the first lobe and the basal half of the second lobe white. Markings dark brown: a costal spot near the base of the cleft and in the centre of the first lobe; three small spots at the terminal half of the dorsum of the first lobe; the terminal half of the second lobe black-brown; and a dark spot centrally in the dorsal fringes of the second lobe.

MALE GENITALIA.– Left valve rounded, with a saccular thorn half as long as the valve and angulated at middle. Right valve more elongated in shape, with a wide saccular process with plough-shaped apex.

FEMALE GENITALIA.– Antrum slightly excavated. Ductus bursae curved, prominent, with numerous longitudinal sclerotised ridges. Bursa copulatrix vesicular, with a pair of longitudinal signa.

DISTRIBUTION.– Canary Islands.

BIOLOGY.– The moth flies in April and May. The hostplant is *Bystropogon plumosus* L'Heritier. The leaves and flower-buds of the central shoots are spun together. Pupation takes place along a stem of the plant.

Wheeleria Tutt, 1905

Wheeleria Tutt, 1905: 37.
  Type species: *Pterophorus spilodactylus* Curtis, 1827; original designation.

DIAGNOSIS.– Forewing with faint to strong markings of white-grey colour. Venation with a single radial vein (R4) present.

MALE GENITALIA.– Valvae asymmetrical. Also the saccular processes are asymmetrical, the process in the right valve being in general smaller than that in the left valve. Aedeagus simple.

FEMALE GENITALIA.– Ostium weakly to strongly excavated. Antrum gradually progressing into the ductus bursae. Bursa copulatrix vesicular, without a signum. Ductus seminalis originates at the junction between the ductus bursae and the bursa copulatrix.

DISTRIBUTION.– Throughout the southern half of the area, the number of species increasing towards the south-eastern corner of the region.

BIOLOGY.– The hostplants are Lamiaceae.

131. *Wheeleria phlomidis* (Staudinger, 1870)

Aciptilus phlomidis Staudinger, 1870: 282.   Pl. 14: 3. Figs 145, 281.

DIAGNOSIS.– Wingspan 25-30 mm. Forewing chalk-white, with a pale brown costal line, darkest at the level of the base of the cleft. Hindwing dark grey-brown, contrasting strongly with the forewing.

MALE GENITALIA.– Left valve rounded with a large saccular spine, as long as the valve. Right valve smaller than the left valve, saccular process with a short blunt apex.

FEMALE GENITALIA.– Ostium saucer-like, centrally excavated. Antrum almost squarish, short. Ductus bursae and bursa copulatrix vesicular, without a signum or sclerotised parts. Apophyses anteriores about as long as papillae anales. Apophyses posteriores three and a half times as long as papillae anales, apically slightly club-like.

DISTRIBUTION.– Greece, as the most western location. Russia, Asia Minor, Iran.

BIOLOGY.– The moth flies in June. The hostplant is *Phlomis* sp. (Staudinger, 1870); Arenberger (1981) mentions *Phlomis orientalis* L.

132. *Wheeleria raphiodactyla* (Rebel, 1900)

Aciptilia raphiodactyla Rebel, 1900: 188.   Pl. 14: 4. Figs 146, 282.

DIAGNOSIS.– Wingspan 21-24 mm. Colour white, tinged greyish, without markings.

Male genitalia.– Valvae slightly asymmetrical. Left valve with a stout and long saccular process, extending beyond the tip of the valve. Right valve has a stout, but shorter, slightly curved spine. Tegumen long and slender. Aedeagus tube-like, short.

FEMALE GENITALIA.– Ostium hardly sclerotised, flat. Antrum a little wider than long. Ductus bursae and bursa copulatrix vesicular, hardly sclerotised. No signum. Apophyses anteriores absent. Apophyses posteriores two and a half times as long as the papillae anales.

DISTRIBUTION.– Spain, France.

BIOLOGY.– The moth flies in July and August. The hostplant is unknown. Moths were flying locally in localities where a limestone formation came to the surface in otherwise granite conditions.

133. *Wheeleria spilodactylus* (Curtis, 1827)

Pterophorus spilodactylus Curtis, 1827: 161.   Pl. 14: 5. Figs 15A, 147, 283.
Aciptilus confusus Herrich-Schäffer, 1855: 384.

DIAGNOSIS.– Wingspan 18-26 mm. Colour yellow-white to white, greyish tinged. Markings consist of grey-brown scale groups: at the base; along the basal half of the costa; at the dorsal margin of the base of the cleft; at the costa just beyond the base of the cleft; and in the top of the second lobe. Fringes white with grey hair-brushes: two at the dorsum of the first lobe; centrally at the costa of the second lobe, and two at the dorsum of the second lobe.

VARIATION.– This species and the following, *W. obsoletus*, show a considerable variation in the intensity of the black markings on the forewing and the colour of the hindwing. The species are differentiated by the black fringe hairs of the anal region of the second lobe of the forewing. In *W. spilodactyla* these hairs reach as far as the total fringe length. In *W. obsoletus* however, the black colour only extends for half the length of the fringe hairs. See figures.

MALE GENITALIA.– Left valve with a centrally placed, hooked saccular process. In the right valve a similar process, which is a little smaller.

FEMALE GENITALIA.– Ostium excavated in a V-shape.

DISTRIBUTION.– Southern half of West and Central Europe and the Mediterranean area, extending into Asia Minor and North Africa.

Continued on p. 145.

The specimens have been collected by and are present in the collection of the author, unless otherwise stated in the captions. All specimens are shown 2.5 x natural size.

PLATE 1

1. *Agdistis tamaricis* (Zeller) – Greece, Peloponnesos, 10 km SW of Nafplion, 0-200 m, 5.viii.1987 (W.O. de Prins).

2. *A. intermedia* Caradja – Hungary, Hortobagy Nat. Park, E of Nagyvan, 10-11.vii.1983 (M. & E. Arenberger).

3. *A. bennetii* (Curtis) – Denmark, SZ, Agersø, e.l. 5.vi.1978 (O. Karsholt).

4. *A. meridionalis* (Zeller) – Great Britain, Dorset, Portland, e.l. 27-29.vii.1988 (O. Karsholt).

5. *A. salsolae* Walsingham – Canary Islands, Fuertaventura, Corralejo, 3-8.iii.1985 (A. Cox).

6. *A. delicatulella* Chrétien – Malta, Zebbug, 6.xi.1983 (P. Sammut).

7. *A. neglecta* Arenberger – Spain, Alicante, Dolores, 24.iv.1981.

8. *A. protai* Arenberger – Turkey, sand dunes at Agköl, S of Silifke (E. Arenberger).

9. *A. adactyla* (Hübner) – Italy, S Tirol, Naturns, 1.vii.1984.

PLATE 2

1. *Agdistis heydeni* (Zeller) – Spain, Barcelona, 15 km SW of Vic, Collsuspina, 900 m, 23-24.vii.1992 (W. Hock).

2. *A. satanas* Millière – France, Les Landes, Lubbon, 5.viii.1986.

3. *A. frankeniae* (Zeller) – Italy, Toscana, Marina di Albarese, Maremma, 19.vi.1981 (M. & E. Arenberger).

4. *A. gittia* Arenberger, paratype – Spain, Granada, Baza, 110 km NE of Granada, 9-10.v.1977 (M. & W. Glaser).

5. *A. espunae* Arenberger, paratype – Spain, Murcia, Alhama de Murcia, Sierra Espuria, 9.x.1975 (M & W. Glaser).

6. *A. glaseri* Arenberger – Spain, Granada, Baza, 110 km NE of Granada, 9-10.v.1977 (M. &W. Glaser).

7. *A. bigoti* Arenberger – Greece, Crete, Sithia Kolpos, Lasithi, 6.xi.1984 (H. Teunissen).

8. *A. symmetrica* Amsel – Tunisia, 20 km W of El Kef, 18-19.vii.1979 (M. & E. Arenberger).

PLATE 3

1. *Agdistis manicata* Staudinger – France, Aude, Île St Lucie, 24.iv.1989 (J. Nel).

2. *A. paralia* (Zeller) – France, Camarque, 28.viii.1961 (L. Bigot).

3. *A. bifurcatus* Agenjo – Spain, Andalusia, near Huelva, 6-8.ix.1961 (G. & W. von Budden-brock).

4. *A. pseudocanariensis* Arenberger – Canary Islands, Tenerife, El Abrigo, 0-20 m, 30.xi.1988 (M. & E. Arenberger).

5. *A. hartigi* Arenberger – Spain, Almeria, 6 km SW of Tabernas, 400 m, 30.ix.1993 (H.W. van der Wolf).

6. *A. betica* Arenberger – Spain, Granada, Cullar de Baza, 28.v.1992 (M. Hull), gent. MH 2846.

7. *Platyptilia tesseradactyla* (Linnaeus) – Canada, Quebec, Aylmer, 27.v.1919 (McDunnough).

8. *P. farfarellus* Zeller – Turkey, 10 km E of Nevsehir, 1300 m, 21.vi.1979 (Groß).

9. *P. nemoralis* Zeller – Slovakia, Murán, 1000 m, 16-17.vii.1990 (H.W. van der Wolf).

10. *P. gonodactyla* ([Denis & Schiffermüller]) – Italy, S Tirol, Gomagoi, 30.vi.1984.

6

7

8

9

10

PLATE 4

1. *Platyptilia calodactyla* ([Denis & Schiffermüller]) – Netherlands, Gelderland, Twello, 11.vii.1989 (J.B. Wolschrijn).

2. Same – Denmark, LFM, Mellemskoven, 30.vi.1992 (O. Karsholt).

3. *P. iberica* Rebel – Spain, Avila, Hoyos del Espino, 12-18.vii.1992.

4. *P. isodactylus* (Zeller) – Netherlands, Zuid Holland, Leiden, xii.1991 (K. Vrieling).

5. *Gillmeria miantodactylus* (Zeller) – Greece, Macedonia, Thessalia, Olympos, 700-2100 m, 21-26.v.1990 (ZMUC).

6. *G. pallidactyla* (Haworth) – France, Pyr. Or., Carol, 1.vii.1976.

7. *G. tetradactyla* (Linnaeus) – Denmark, Sillestrup Strand, 13.vii.1955 (E. Pyndt).

8. *Lantanophaga pusillidactylus* (Walker) – Ecuador, Galapagos Archipelago, Isabela Island, nr Tagus Cove, 100 m, 21.v.1992 (B. Landry).

9. *Stenoptilodes taprobanes* (Felder & Rogenhofer) – W Turkey, Anamur, 0-50 m, 8-12.v.1991 (A. Cox).

10. *Paraplatyptilia metzneri* (Zeller) (27 mm) – France, Savoie, 2 km N of Col de Galibier, 2350 m, 1-3.viii.1991.

PLATE 5

1. *Amblyptilia acanthodactyla* (Hübner) – Denmark, NEJ, Sandmilen, 24.vii.1981 (O. Karsholt).

2. *A. punctidactyla* (Haworth) – Austria, Salzburg, Gerlos Pass, 7.viii.1984.

3. *Stenoptilia graphodactyla* (Treitschke) – Austria, N Tirol, Mühlauer Klamm, 30.vi.1966 (Hornegger).

4. *S. pneumonanthes* (Büttner) – Denmark, NEJ, Læsø, 7.vii.1982 (O. Karsholt).

5. *S. gratiolae* Gibeaux & Nel, paratype – France, env. Mouy/S. Neuvry, 15.vii.1989 (C. Gibeaux).

6. *S. pterodactyla* (Linnaeus) – France, Alpes de Hte Provence, St Anne-Condaminé, 1500-1750 m, 24-27.vii.1987.

7. *S. mannii* (Zeller) – Turkey, Kireçli Geçidi, W side, 20 km E of Tortum, 2100-2400 m, 9.vii.1989 (W.O. de Prins).

8. *S. veronicae* Karvonen – Finland, Inari, 18.vii.1985 (G. Langohr).

9. *S. bipunctidactyla* (Scopoli) – Netherlands, Noord Holland, Egmond aan Zee, 22.vii.1994.

10. *S. aridus* Zeller – Spain, Teruel, Valdeltormo, 23.vi.1982.

6

7

8

9

10

PLATE 6

1. *Stenoptilia elkefi* Arenberger – Turkey, Van Gölü, 1800 m, 25.vi.1965 (H. Noack).

2. *S. lucasi* Arenberger, paratype. – Turkey, Develi, Erciyesdagh, 1700 m, 11-18.vii.1970 (Friedel).

3. *S. annadactyla* Sutter – France, Seine et Marne, Forêt de Fontainebleau, e.l. 18.vii.1989 (Ch. Gibeaux).

4. *S. pelidnodactyla* (Stein) – Sweden, Öl., Högsrum, 5.vi.1982 (O. Karsholt).

5. *S. reisseri* Rebel – Spain, Sierra de Gredos, Garganta de las Pozas, 1800 m, 11-22.vii.1980 (M. & E. Arenberger).

6. *S. hahni* Arenberger – Spain, Granada, Sierra Nevada, Veleta Road, 2000 m, 23-29.vii.1986.

7. *S. millieridactyla* (Bruand) – Great Britain, Holloway Derbys, la. 18.vi.1966 (D. Agassiz).

8. *S. islandicus* (Staudinger) – Norway, STi, Kongsvoll, la. 12.vi.1985 (O. Karsholt).

9. *S. parnasia* Arenberger – Greece, Mount Parnass, western slope, 1800 m, 2.viii.1977 (Groß), gent. CG 2334.

10. *S. coprodactylus* (Stainton) – France, Savoie, 8 km N of Col de Galibier, 2000 m, 27.vii.1988.

11. *S. lutescens* (Herrich-Schäffer) – France, Alpes de Hte Provence, Lautaret, 2050 m, 12.vii.1988 (Ch. Gibeaux).

7

8

9

10

11

PLATE 7

1. *Stenoptilia nepetellae* Bigot & Picard – France, Alpes Mar., Col de Couillole, 1600 m, 19.vii.1987.

2. *S. stigmatodactylus* (Zeller) – Greece, Macedonia, Aposkepos Vernon, 1000 m, 6.vii.1981 (W.O. de Prins).

3. *S. stigmatoides* Sutter & Skyva – Slovakia, Medovarce, 18.vi.1994 (J. Skyva).

4. *S. zophodactylus* (Duponchel) – Spain, Teruel, Mora de Rubielos, 24.vi.1976.

5. *Buszkoiana capnodactylus* (Zeller) – Denmark, SZ, Lellinge Skov, e.p. 7.vii.1958 (P.K. Nielsen).

6. Same – Denmark, NEZ, Jægersborg, la. 20.vi.1954 (N.L. Wolff).

7. *Cnaemidophorus rhododactyla* ([Denis & Schiffermüller]) – Spain, Huesca, Linas de Broto, 6.vii.1982.

8. *Marasmarcha lunaedactyla* (Haworth) – Sweden, Öl., Högsrum, larva 5.vi.1982 (O. Karsholt).

9. *M. fauna* (Millière) – France, Drôme, La Penne sur Ouvèze, 9-15.vii.1991.

10. *M. oxydactylus* (Staudinger) – France, Alpes de Hte Provence, St Paul sur Ubaye, 1400-1800 m, 20-26.vii.1987.

6

7

8

9

10

PLATE 8

1. *Geina didactyla* (Linnaeus) – France, Pyr. Or., Carol, 8.vii.1985.

2. *Procapperia maculatus* (Constant) – France, Alpes de Hte Provence, La Condaminé - Chatelard, 1300-1800m, 20-28.vii.1987.

3. *Paracapperia anatolicus* (Caradja) – W Iran, Kurdestan, Road Baneh- Marivan, 86 km SE of Baneh, 1950 m, 5.vii.1975 (Ebert & Falkner).

4. *Capperia britanniodactylus* (Gregson) – Netherlands, Gelderland, Terlet, e. l. 22.vi.1994.

5. *C. celeusi* (Frey) – Turkey, Gümüshane, Tersundagi Geçidi, 2000 m, 24.vii.1987 (W. van Oorschot).

6. *C. trichodactyla* ([Denis & Schiffermüller]) – Poland, Torun, e. l. 14.vi.1977 (J. Buszko).

7. *C. fusca* (Hofmann) – France, St Rémy de Blot, 500 m, e. l. 31.vii.1987 (J. Nel).

8. *C. bonneaui* Bigot – Spain, Teruel, Albarracin, Valdevecar, 1200 m, 30.vii.1991 (H.W. van der Wolf), gent. CG 6307.

9. *C. hellenica* Adamczewski – Spain, Teruel, Fornoles, 25 km SE of Alcaniz, 24.vii.1992.

10. *C. loranus* (Fuchs) – Slovakia, Zadiel, 18-20.vii.1990 (H.W. van der Wolf).

11. *C. marginellus* (Zeller) – Italy, Sicily, no date (NNM).

6

7

8

9

10

11

PLATE 9

1. *Capparia zelleri* Adamczewski – Italy, Sicily, Catan, 4.vii, gent. BM 5402 (BMNH).

2. *C. polonica* Adamczewski – France, Bouches-du-Rhône, La Ciotat, e. l. 19.iii.1985 (J. Nel).

3. *C. maratonica* Adamczewski – France, Bouches-du-Rhône, Etang des Aulmes, 26.vii.1986 (L. de Ridder).

4. *Buckleria paludum* (Zeller) – Netherlands, Overijssel, Losser, 7.viii.1992.

5. *Oxyptilus pilosellae* (Zeller) – France, Isère, Freney d'Oisans, 1000-1350 m, 25.vii-5.viii.1988.

6. *O. chrysodactyla* ([Denis & Schiffermüller]) – Belgium, Brabant, Jette, 90 m, 21.vi.1976 (F. Coenen).

7. *O. ericetorum* (Stainton) – Denmark, NEZ, Asserbo, 3.viii.1981 (K. Schnack).

8. *O. parvidactyla* (Haworth) – Germany, Rheinland, Kaub, 25.v.1985.

9. Same – f. *hoffmannseggi* Möschler – Spain, Avila, Hoyos del Espino, 12-18.vii.1992.

10. *Crombrugghia distans* (Zeller) – Austria, Niederösterreich, Neu Aigen, 24.v.1986.

11. *C. tristis* (Zeller) – Spain, Cordoba, Venta de Azuel, 17.v.1981.

12. Same – Poland, Piotrkow Tr., Lubiaszow, 7.vii.1974 (J. Buszko).

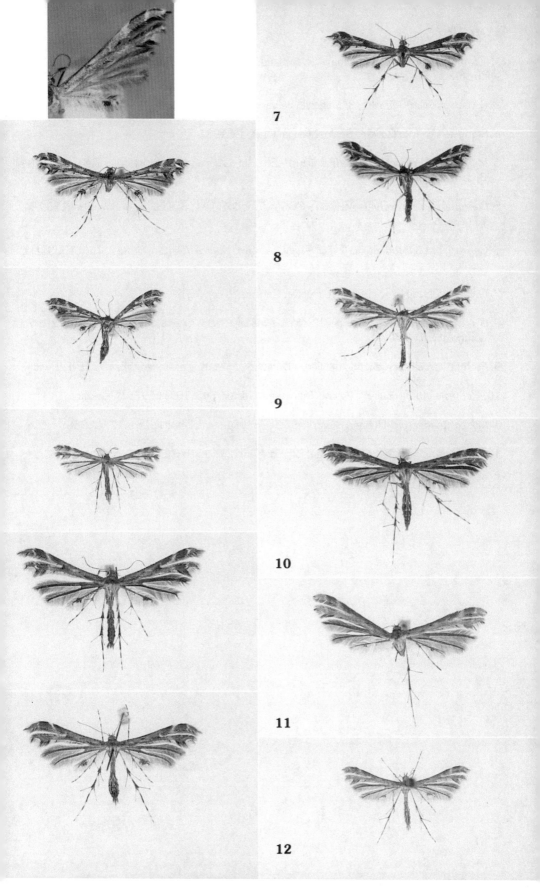

7

8

9

10

11

12

PLATE 10

1. *Crombrugghia kollari* (Stainton) – Switzerland, Wallis, Ausserberg, nr Visp, 700 m, 13.vii.1968 (Groß).

2. *C. laetus* (Zeller) – France, St Marcellin de Vars, 1650 m, e. l. 12.vii.1986 (J. Nel).

3. *Stangeia siceliota* (Zeller) – Spain, Malaga, Ojen, 4.v.1981.

4. *Megalorhipida leucodactylus* (Fabricius) – Ecuador, Galapagos Archipelago, Isabela, 5.iii.1989 (B. Landry).

5. *Puerphorus olbiadactylus* (Millière) – Cyprus, Troodos Mtns., S of Lania, 600 m, 29.x.1989 (M. & E. Arenberger).

6. *Gypsochares baptodactylus* (Zeller) – Italy, Sardinia, Nuoro, Villanova-Strisaili, 7.viii.1983 (J.H. Kuchlein).

7. *G. bigoti* Gibeaux & Nel – Spain, Huelva, Calañas, 15.v.1981.

8. *G. nielswolffi* Gielis & Arenberger, holotype. – Madeira, Serra d'Agua, Station Salazar, 6-7.ix.1973 (Lomholdt & Wolff).

9. *Pselnophorus heterodactyla* (Müller) – Denmark, LFM, Kosterskoven, 12.vi.1975 (E. Pyndt).

10. *Hellinsia inulae* (Zeller) – Poland, Inowroctuw Mstuy, pup. 10.viii.1973 (J. Buszko).

11. *H. carphodactyla* (Hübner) – Germany, Bayern, Fürstenfeldbruck, 16.v-12.vi.1986.

12. *H. chrysocomae* (Ragonot) – Great Britain, Kent, 21.vii.1970 (E. S. Bradford).

7

8

9

10

11

12

PLATE 11

1. *Hellinsia osteodactylus* (Zeller) – Germany, Bayern, Kaiserwacht, 9.vii.1984.

2. *H. pectodactylus* (Staudinger) – Spain, Canary Islands, Tenerife, Chio, 5 km SE at road C-823, 900 m, 12.iv.1993 (H.W. van der Wolf).

3. *H. distinctus* (Herrich-Schäffer) – Denmark, NEZ, Lundtofte, 1.viii.1978 (K. Schnack).

4. *H. didactylites* (Ström) – Switzerland, Graubünden, Löbbia, 28.vi.1984.

5. *H. tephradactyla* (Hübner) – Switzerland, Wallis, Laquintal, Simplon, 1500 m, 5-15.vii.1955 (Groß).

6. *H. lienigianus* (Zeller) – Netherlands, Noord Holland, Overveen, 19.vi.1994.

7. *Oidaematophorus lithodactyla* (Treitschke) – Germany, Rheinland, Kaub, 30.vi.1985.

8. *O. rogenhoferi* (Mann) – Norway, Sti, Kongsvoll, 900-1100 m, 20-28.vii.1983 (O. Karsholt).

9. *O. constanti* (Ragonot) – Slovakia, Krupinská, Medovarce, 5.vi.1979 (J. Marek).

10. *O. giganteus* (Mann) – France, Var, Hyères, Ile du Levant, 11.v.1989 (J. Nel).

7

8

9

10

PLATE 12

1. *Oidaematophorus vafradactylus* Svensson – Sweden, Öland, Hulterstad, la. 29.v.1981 (K. Schnack).

2. *Emmelina monodactyla* (Linnaeus) – Denmark, SZ, Strøby Jerne, 11.ix.1955 (E. Traugott-Olsen).

3. Same – Denmark, LFM, Blans, 5.ix.1971 (E. Pyndt).

4. *E. argoteles* (Meyrick) – Germany, Hessen, Torfgrube S of Pfungstadt, 26.vi.1970 (Groß).

5. *Adaina microdactyla* (Hübner) – Denmark, SZ, Holmegårds Mose, la. 6.v.1979 (O. Karsholt).

6. *Calyciphora punctinervis* (Constant) – Spain, Gerona, Darnius, 150 m, 22.ix.1984 (F. Coenen).

7. *C. homoiodactyla* (Kasy) – Greece, Lithochoron Plaka, 29.v.1969 (Lukasch).

8. *C. adamas* (Constant) – Spain, Granada, Pto de Mora, 21.ix.1979.

9. *C. acarnella* (Walsingham) – Italy, Sardinia, Aritzo, 1100 m, 16.viii.1975 (F. Hartig).

PLATE 13

1. *Calyciphora albodactylus* (Fabricius) – Spain, Navarra, Sada, 23.vii.1985.

2. *C. xanthodactyla* (Treitschke) – Austria, Hundsheim, 9.viii.1978 (E. Arenberger).

3. *C. nephelodactyla* (Eversmann) – France, Alpes de Hte Prov., 10 km E of S Paul, Maurin, 1750-1800 m, 27-28.vii.1991.

4. *Porrittia galactodactyla* ([Denis & Schiffermüller]) – Netherlands, Noord Holland, Overveen, 15-26.vi.1986.

5. *Merrifieldia leucodactyla* ([Denis & Schiffermüller]) – Denmark, NWZ, Bjergsted Bakker, 12.vii.1979 (O. Karsholt).

6. *M. tridactyla* (Linnaeus) – France, Corsica, Corté, Restonica, 550 m, e. l. 26.iv.1988 (J. Nel).

7. *M. baliodactylus* (Zeller) – Germany, Bayern, Kaiserwacht, 9.vii.1984.

8. *M. malacodactylus* (Zeller) – Spain, Murcia, Aledo, 20.ix.1979.

9. *M. semiodactylus* (Mann) – France, Corsica, Ajaccio, St Antone, 80 m, 30.iv.1988 (J. Nel).

10. *M. hedemanni* (Rebel) – Canary Islands, Tenerife, Chio, 5 km SE road C-823, 900 m, 12.iv.1993 (H.W. van der Wolf).

5

6

7

8

9

10

PLATE 14

1. *Merrifieldia chordodactylus* (Staudinger) – Spain, Malaga, Periana, 24.iv.1978.

2. *M. bystropogonis* (Walsingham) – Canary Islands, Tenerife, Chio, 5 km SE road C-823, 900 m, 12.iv.1993 (H.W. van der Wolf).

3. *Wheeleria phlomidis* (Staudinger) – Turkey, Konya, Sultan Daglari, Aksehir Pass, 1700 m, 4.vi.1974 (Groß).

4. *W. raphiodactyla* (Rebel) – Spain, Granada, Sierra Nevada, Veleta Road, 2000 m, 14.vii.1971 (E. Arenberger), gent. CG 2381.

5. *W. spilodactylus* (Curtis) – Great Britain, Wales, Gwynned, 6.vii.1984 (M. Hull).

6. *W. obsoletus* (Zeller) – "Yugoslavia", Dalmatia, Bast, 13-27.vi.1990 (J.B. Wolschrijn).

7. *W. lyrae* (Arenberger), paratype – Greece, Lakonia, waterfall at Nomia-Lyra, 17.v.1979 (Gozmány & Christensen).

8. *W. ivae* (Kasy) – Lebanon, mountains, cedars nr Becharré, 1900-2000 m, 3-5.vi.1969 (Groß).

9. *Pterophorus pentadactyla* (Linnaeus) – Denmark, LFM, Mellemskoven, 22.vii.1942 (W. van Deurs).

10. *P. ischnodactyla* (Treitschke) – France, Basses Alpes, St Michel l'Observatoire, e. l. 21.iv.1986 (J. Nel).

6

7

8

9

10

PLATE 15

1, 2. Larva and pupa of *Agdistis bennetii* (Curtis).

3. Larva of *Platyptilia gonodactyla* ([Denis & Schiffermüller]).

4. Larva of *Amblyptilia acanthodactyla* (Hübner).

5, 6. Larva and pupa (green form) of *Stenoptilia nepetellae* Bigot & Picard.

7, 8. Larva and pupa of *Marasmarcha fauna* (Millière).

9-11. Larva and pupae of *Stenoptilia zophodactylus* (Duponchel).

1

2

3

4

5

6

7

8

9

10

11

PLATE *16*

1-3. Larva and pupae of *Capperia britanniodactylus* (Gregson).

4, 5. Pupae of *Oidaematophorus lithodactyla* (Treitschke).

6, 7. Pupae of *Merrifieldia baliodactylus* (Zeller).

8. Larva (4th instar) of *Hellinsia lienigianus* (Zeller).

9. Larva of *Porrittia galactodactyla* ([Denis & Schiffermüller]).

10. Larva of *Pterophorus pentadactyla* (Linnaeus).

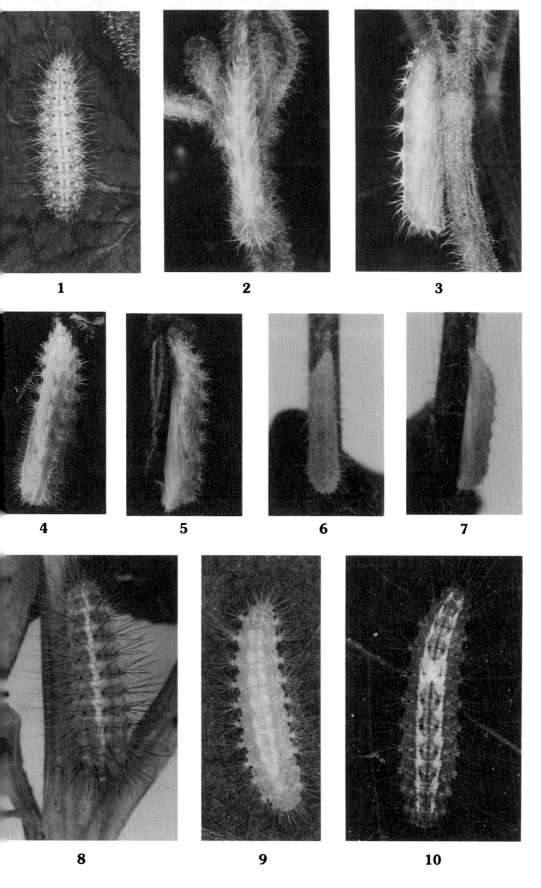

1

2

3

4

5

6

7

8

9

10

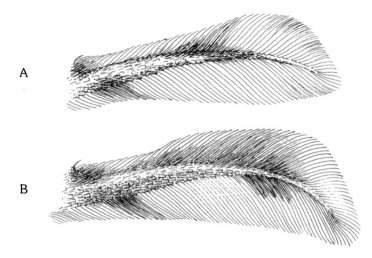

Fig. 15. Second forewing lobe of A, *Wheeleria spilodactylus* and B,
W. obsoletus. (R. Nielsen del.).

BIOLOGY.– The moth flies from April to October, depending on the latitude. In the Mediterranean area probably in two or three generations. The hostplant is *Marrubium
vulgare* L. (South, 1881; Amsel, 1935b; Beirne, 1954; Hannemann, 1977b; Buszko,
1986; Michaelis, 1986; Nel, 1989d; Gieliš, bred). Also recorded is *Ballota nigra* L.
(Mitterberger, 1912; Hannemann, 1977b; Bigot & Picard, 1983).

134. Wheeleria obsoletus (Zeller, 1841)

Pterophorus obsoletus Zeller, 1841: 859.
Pterophorus desertorum Zeller, 1867: 386.
Alucita gonoscia Meyrick, 1922: 549.
Alucita phlomidactylus Wasserthal, 1970: 213.
Alucita marrubii Wasserthal, 1970: 214.

Pl. 14: 6. Figs 15B, 148, 284.

DIAGNOSIS.– Wingspan 18-22 mm. Colour pale white-yellow, speckled with sparse pale
brown scales. In the fringes at the costa and the dorsum just beyond the base of the
cleft a brown-black hair-brush; at the dorsum before the apex of both lobes a brown
margin of hairs.

MALE GENITALIA.– Left valve slightly larger than right valve. In the left valve a curved,
rather stout saccular spine, half as long as the valve. In the right valve a short stout,
hooked saccular spine.

FEMALE GENITALIA.– Ostium excavated in a U-shape. Antrum as long as ductus bursae,
membraneous and slightly more sclerotised.

DISTRIBUTION.– Czechoslovakia, Austria, Italy, France, Poland, the Balkan countries, Greece,
and extending into Asia Minor.

BIOLOGY.– The moth flies from April till August. The hostplants are *Phlomis cretica* Pres.
and *Marrubium peregrinum* L. (Marek & Skyva, 1985).

135. *Wheeleria lyrae* (Arenberger, 1983)

Pterophorus lyrae Arenberger, 1983: 204.   Pl. 14: 7. Figs 149, 285.

DIAGNOSIS.– Wingspan 17-18 mm. Colour white. In the forewing a brown line extends along the costa from the base up to one third of the costa on the first lobe. Fringes white with dark dashes as follows: at middle of the dorsum of the first lobe; at two thirds of the costa of the second lobe; at the dorsum of the wing at the base of the cleft, and near the apex of the second lobe. First and second hindwing lobes brown, third lobe white.

MALE GENITALIA.– Left valve with a short, stout, hooked saccular process. The right valve more lanceolate than left valve, and with a short, strong, straight saccular process, parallel to the valve margin.

FEMALE GENITALIA.– Ostium slightly excavated (in *W. spilodactyla* excavated in a V-shape, in *W. obsoletus* in a U-shape). Antrum as long as wide.

DISTRIBUTION.– Greece.

BIOLOGY.– The moth flies in May. The hostplant is unkown.

136. *Wheeleria ivae* (Kasy, 1960)

Aciptilia ivae Kasy, 1960: 183.   Pl. 14: 8. Figs 150, 286.

DIAGNOSIS.– Wingspan 20-25 mm. Colour grey-white, speckled with grey-brown scales. Terminal half of first lobe, and second lobe, white. Brown scaling extends along the basal half of the costa, near the wing base and as an extension towards the wing base from the second lobe. Fringes white, except for the terminal fourth of the dorsum of the first lobe and along the second lobe slightly further toward the base.

MALE GENITALIA.– Left valve lanceolate, with a strong, hooked saccular process. Right valve of about the same shape as the left valve, the saccular process being more slender.

FEMALE GENITALIA.– Ostium simple, small. Antrum hardly sclerotised, with some longitudinal sclerotised ridges. Bursa copulatrix vesicular, membraneous. Apophyses anteriores well developed.

DISTRIBUTION.– Ex-Yugoslavia, extending into Asia Minor, Syria and Lebanon.

BIOLOGY.– The moth flies from the end of May through the month of June. The hostplant is *Stachys iva* Griseb., growing in rocky valleys and slopes. The plant seems to be rare and local. The larva feeds in the felty central part of the plant.

Pterophorus Schäffer, 1766

Pterophorus Schäffer, 1766: t. 104, figs 2, 3.
  Type species: *Phalaena Alucita pentadactyla* Linnaeus, 1758; subsequent designation (Whalley, 1961).
Pterophorus Geoffroy, 1762; suppressed (ICZN Op. 228).
Plumiger Valmont-Bomare, 1791; unavailable (ICZN Op. 228).

Pterophora Hübner, [1806]; suppressed (ICZN Op. 97).
Pterophora Hübner, 1822: 80, 81.
  Type species: *Phalaena pentadactyla* Linnaeus, 1758; subsequent designation (Tutt, 1905).
Aciptilia Hübner, [1825]: 430.
  Type species: *Phalaena pentadactyla* Linnaeus, 1758; subsequent designation (Tutt, 1905).
Aciptilus Zeller, 1841; emendation.
Acoptilia Agassiz, 1847; emendation.
Acoptilus Agassiz, 1847; emendation.
Alucita auct., (*nec* Linnaeus, 1758) (ICZN Op. 703).

DIAGNOSIS.– Forewing cleft from middle or less. Lobes acute, without a terminal margin. R1 absent, R4 present, M3 and Cu2 forked or fused. Hindwing without a scale-tooth.

MALE GENITALIA.– Valvae asymmetrical, with asymmetrical processes. At the base of the valvae well developed large scaly hairs.

FEMALE GENITALIA.– Bursa copulatrix blister-like, without or with one or two signa; vesica seminalis often well developed.

DISTRIBUTION.– Known from all regions. The main distribution lies in the southern parts of the Palaearctic region. In Africa about ten species are known, in the Indo-Australian fauna thirty, and in the New World four species.

BIOLOGY.– The hostplants belong to the Convolvulaceae.

137. *Pterophorus pentadactyla* (Linnaeus, 1758)

Phalaena Alucita pentadactyla Linnaeus, 1758: 542.   Pl. 14: 9; pl. 16: 10. Figs 6, 151, 287.
Phalaena tridactyla Scopoli, 1763: 257.

DIAGNOSIS.– Wingspan 24-35 mm. Colour silvery white, speckled with sparse black scales.

MALE GENITALIA.– The left valve with a large strongly curved saccular process, two thirds as long as the valve. The right valve with a shorter and more smoothly curved spine.

FEMALE GENITALIA.– Ostium membraneous. Antrum membraneous, almost square. Ductus bursae and ductus seminalis strong, with numerous longitudinal sclerotised ridges. Bursa copulatrix with a rosette-like signum, composed by small spiculae.

DISTRIBUTION.– The whole of the area, except for the extreme south. To the east through Asia Minor into Iran.

BIOLOGY.– The moth flies from May to September. The hostplants are *Convolvulus arvensis* L. (de Graaf, 1859; Hannemann, 1977b; Buszko, 1986; Nel, 1987a) and *Calystegia sepium* (L). (de Graaf, 1859; Hannemann, 1977b; Emmet, 1979; Buszko, 1986; Arenberger & Jaksic, 1991; Gielis, bred). The plants grow in a great variety of surroundings, and are very common. The larva first feeds on the young leaves, later the flower-buds and flowers are eaten. The larva rests on the underside of a leaf, or along a stem. The leaves are eaten from the underside, causing brown spotting. Pupation under a leaf.

138. *Pterophorus ischnodactyla* (Treitschke, 1833)

Alucita ischnodactyla Treitschke, 1833: 223.   Pl. 14: 10. Figs 152, 288.
Aciptilia actinodactyla Chrétien, 1891: 99.
Aciptilia eburnella Amsel, 1968: 14.

DIAGNOSIS.– Wingspan 16-18 mm. Colour greyish cream-white. The species is characterized by the small black dots along the costa and dorsum of the forewing.

MALE GENITALIA.– Left valve with a widely forked costal process; a single hooked process
in the right valve.

FEMALE GENITALIA.– Ostium flat. Antrum almost membraneous, rounded. The distal margin of the 7th tergite smoofhly excavated. Ductus bursae is bulged out, with hardly
any transverse ridges.

DISTRIBUTION.– Southern half of West and Central Europe, the Mediterranean area and
extending into Asia Minor, Syria, Lebanon and Iraq. Also recorded from North and
South Africa.

BIOLOGY.– The moth flies from April to September, in the southern half of the area in
several generations. The hostplant is *Convolvulus cantabricus* L. (Nel, 1987a; Bigot et
al., 1990), which grows on rather dry, submediterranean localities. Occasionally up to
four larvae can be found on one hostplant shoot (Nel, 1987a).

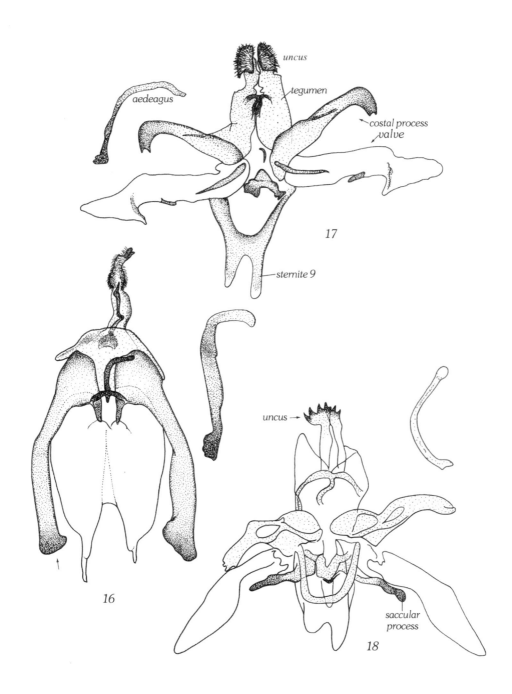

MALE GENITALIA

16. *Agdistis tamaricis* (Zeller) – Tunisia, Nefta, 5-18.v.1988 (Zool. Mus. Exp. Copenhagen), gent. CG 4201 (ZMUC).
17. *A. intermedia* Caradja – Hungary, Hortobagy Nat. Park, E of Nagyvan, 10-11.vii.1983 (M. & E. Arenberger), gent. Arenberger 1211 (Arenberger).
18. *A. bennetii* (Curtis) – Netherlands, Zeeland, Vlissingen, 15.viii.1980, gent. CG 2400.

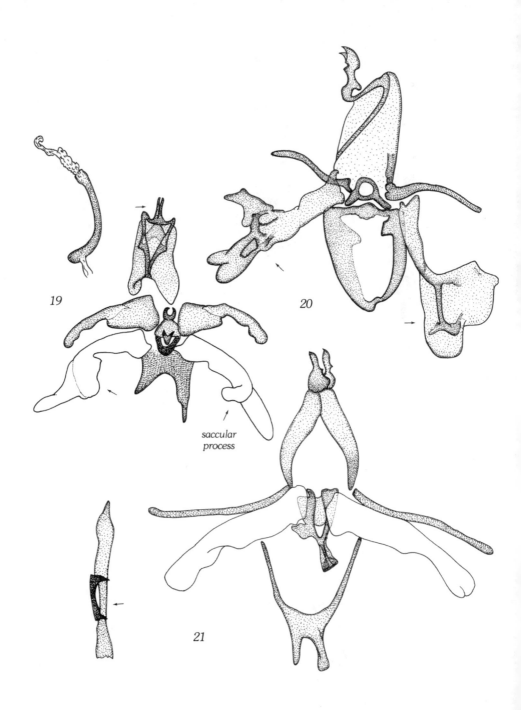

saccular process

MALE GENITALIA

19. *Agdistis meridionalis* (Zeller) – France, Var, Giens, 10.iv.1984 (J. Nel), gent. CG 2372.
20. *A. salsolae* Walsingham – Canary Islands, Tenerife, no date, gent. BM 13307 (BMNH).
21. *A. delicatulella* Chrétien – Malta, Zebbug, 2.xi.1983 (P. Sammut), gent. CG 1543.

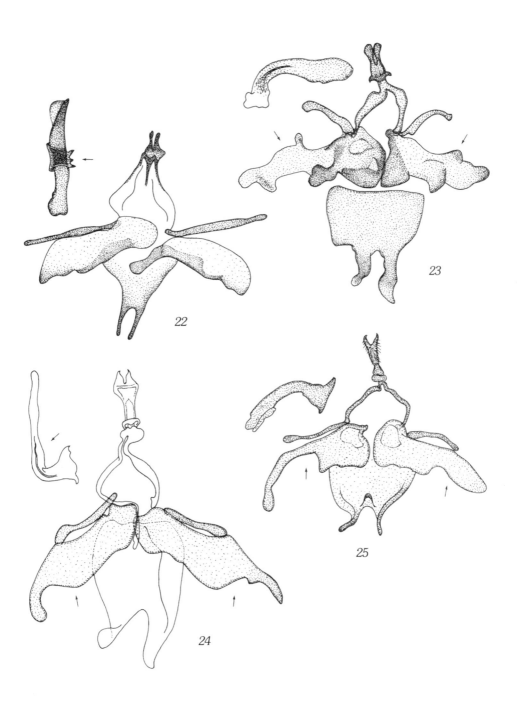

MALE GENITALIA

22. *Agdistis neglecta* Arenberger – France, Cannes, 13.vi.1885, gent. CG 3020 (NNM).
23. *A. protai* Arenberger – Italy, Sardinia, Cagliari, Poetto, 17.vii.1936 (H.G. Amsel), gent. Jäckh 4640 (Kol. Mus. Bremen).
24. *A. adactyla* (Hübner) – Spain, Huesca, Biescas, 1.viii.1989, gent. CG 2417.
25. *A. heydeni* (Zeller) – Spain, Navarra, Sada, 23.vii.1985, gent. CG 2096.

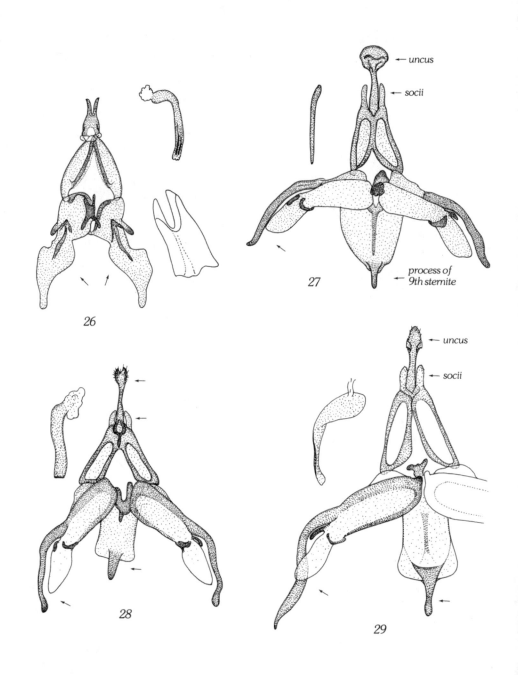

MALE GENITALIA

26. *Agdistis satanas* Millière – Italy, Sardinia, Aritzo, 30.vii-4.viii.1981 (Kaltenbach), gent. CG 2081.

27. *A. frankeniae* (Zeller) – Mauretania, no date, gent. CG 3013 (NNM).

28. *A. gittia* Arenberger – Spain, Granada, Baza, 110 km NE of Granada, 9-10.v.1977 (M. & W. Glaser), gent. Arenberger 2667 (Arenberger).

29. *A. espunae* Arenberger – Spain, Murcia, Alhama de Murcia, 19-20.ix.1974 (M. & W. Glaser), gent. Arenberger 1547 (Glaser).

MALE GENITALIA

30. *Agdistis glaseri* Arenberger – Spain, Murcia, Alhama de Murcia, Sierra Espuna, 8.x.1975 (M. & W. Glaser), gent. Arenberger 454 (Arenberger).
31. *A. bigoti* Arenberger – Greece, Crete, Sithia Kolpos, Lasithi, 6.xi.1984 (H. Teusissen), gent. CG 2211.
32. *A. symmetrica* Amsel – Gent. Arenberger 2014 (Mus. Vind.).

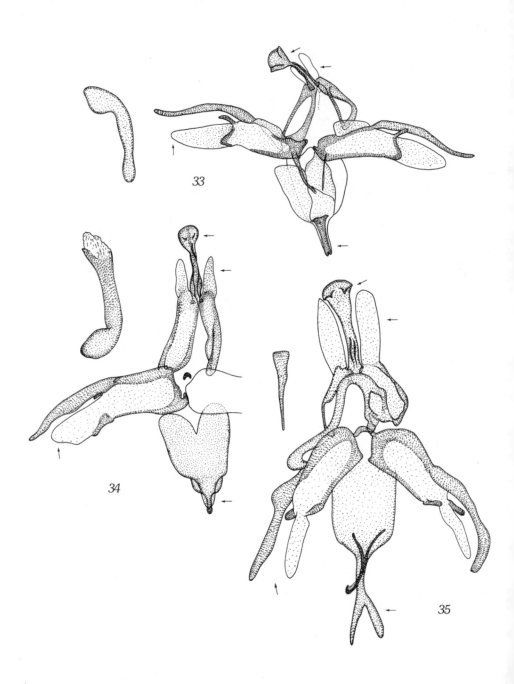

MALE GENITALIA

33. *Agdistis manicata* Staudinger – Spain, Cadiz, Pto Sta Maria, 27.vii.1987, gent. CG 2401.
34. *A. paralia* (Zeller) – Spain, Alicante, Denia, Las Marinas, 5.ix.1981 (J. Koster), gent. CG 2375.
35. *A. bifurcatus* Agenjo – Spain, Cadiz, Tarifa, Rio Jara, 5-6.x.1993 (H.W. van der Wolf), gent. CG 2510.

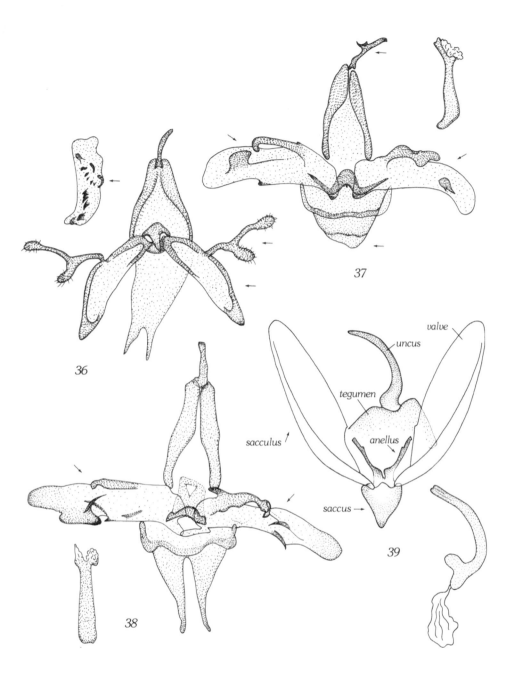

MALE GENITALIA

36. *Agdistis pseudocanariensis* Arenberger – Canary Islands, Fuerteventura, Corralejo, 3-8.iii.1985 (A. Cox), gent. CG 2438.

37. *A. hartigi* Arenberger – Spain, Murcia, Alhama de Murcia, 27-29.ix.1973 (M. & W. Glaser), gent. Arenberger 1542 (Glaser).

38. *A. betica* Arenberger – Spain, Malaga, Road Tarifa-Algeciras, 14 km W of Algeciras, 200 m, 9.vi.1974 (M. & W. Glaser), gent. Arenberger 1377 (Glaser).

39. *Platyptilia tesseradactyla* (Linnaeus) – Canada, British Columbia, Clinton, 13.v.1938 (J.K. Jacob), gent. CG 2323.

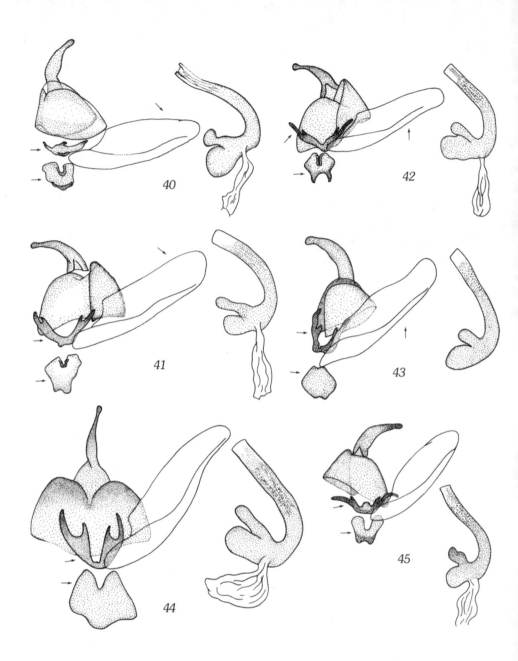

MALE GENITALIA

40. *Platyptilia farfarellus* Zeller – Italy, Garda, Voltino di Tremosine, 9.vi.1983 (J. Neyts), gent. CG 2029.
41. *P. nemoralis* Zeller – Austria, Nieder Österreich, Hohe Wand, 31.vii.1980 (E. Arenberger), gent. CG 2326.
42. *P. gonodactyla* ([Denis & Schiffermüller]) – Belgium, Antwerpen, Niel, 30.viii.1987 (W.O. de Prins), gent. CG 1917 (W.O. de Prins).
43. *P. calodactyla* ([Denis & Schiffermüller]) – France, Drôme, La Penne sur Ouvèze, 15-19.vii.1986 (H.W. van der Wolf), gent. CG 1843.
44. *P. iberica* Rebel – Spain, Avila, Laguna de Gredos, 2100 m, 17.vii.1934 (Reisser), gent. CG 1751 (Mus. Vind.).
45. *P. isodactylus* (Zeller) – England, no date, gent. CG 3126 (NNM).

156

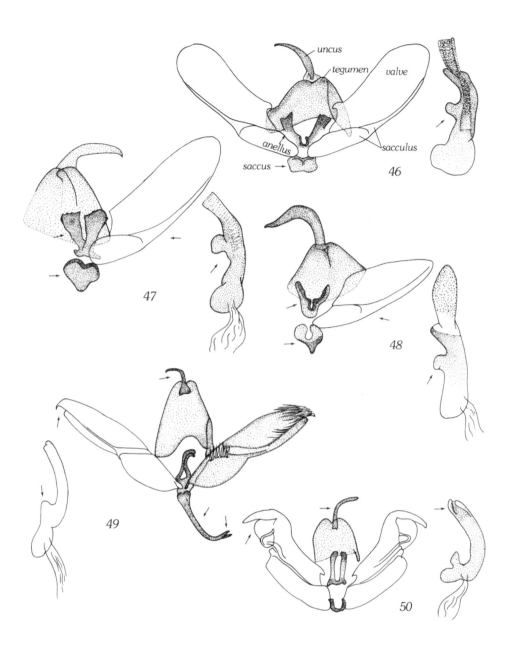

MALE GENITALIA

46. *Gillmeria miantodactylus* (Zeller) – France, Pyr. Or., La Puignal, 1400 m, 17.viii.1988 (Hock), gent. CG 1994 (A. Cox).
47. *G. pallidactyla* (Haworth) – France, Pyr. Or., La Pla, 26 km E Ax-les-Thermes, 1000 m, 26.vii.1988 (R.T.A. Schouten), gent. CG 1995.
48. *G. tetradactyla* (Linnaeus) – France, Reims, 13.vii.1975, gent. CG 2327.
49. *Lantanophaga pusillidactylus* (Walker) – Peru, Lima, Miraflores, 30 m, 19-21.i.1987 (O. Karsholt), gent. CG 4153 (ZMUC).
50. *Stenoptilodes taprobanes* (Felder & Rogenhofer) – Canary Islands, Puerto de la Cruz, 25.iii-5.iv.1968 (van Aartsen), gent. CG 1566 (ITZ).

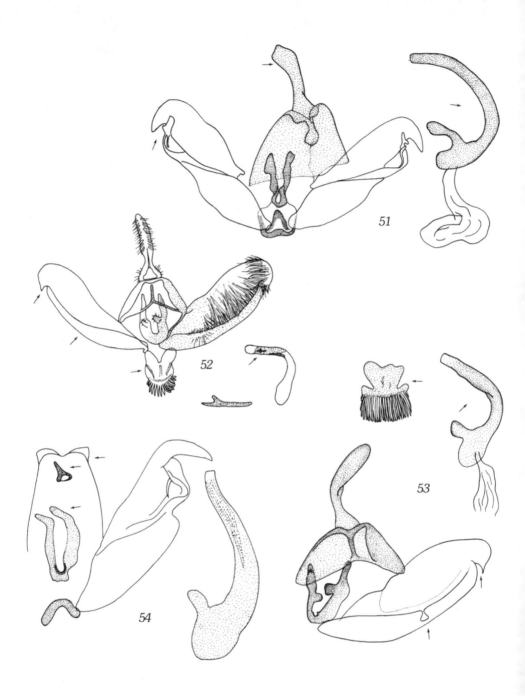

51. *Paraplatyptilia metzneri* (Zeller) – France, Htes Alpes, Les Orres, 7-9.vii.1988 (A. Cox), gent. CG 2403.
52. *Amblyptilia acanthadactyla* (Hübner) – Spain, Cuenca, Cuenca, 29.vi.1982, gent. CG 2321.
53. *A. punctidactyla* (Haworth) – Austria, Tirol, Gerlos Pass, 8.vii.1984, gent. CG 2041.
54. *Stenoptilia graphodactyla* (Treitschke) – Germany, Württemberg, Rauhe Alb, 26.vi.1938 (Grabe), gent. CG 6253 (Löbb. Mus.)

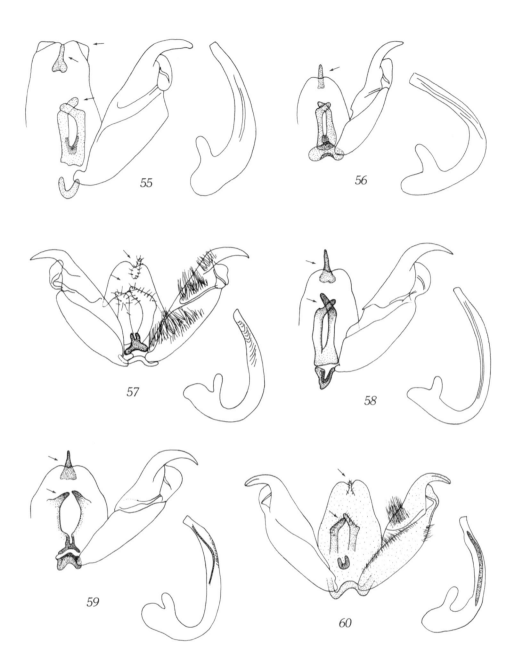

55

56

57

58

59

60

Male Genitalia

55. *Stenoptilia pneumonanthes* (Büttner) – Netherlands, Friesland, Appelscha, 17.viii.1980, gent. CG 1510.

56. *S. gratiolae* Gibeaux & Nel – Austria, Nieder Österreich, Gramnatnen, Fürbachwiesen, 26.vii.1967 (F. Kasy), gent. Mus. Vind. 7517 (Mus. Vind.).

57. *S. pterodactyla* (Linnaeus) – Turkey, Tanin, 25.vi.1985 (C. Zwakhals), gent. CG 2105.

58. *S. mannii* (Zeller) – Turkey, Van, 30-33 km NE Catah, 2000-2400 m, 5.viii.1988 (W.O. de Prins), gent. CG 6016.

59. *S. veronicae* Karvonen – Finland, Lapponia inarensis, Ingri, 18.vii.1985 (G. Langohr), gent. CG 2332.

60. *S. bipunctidactyla* (Scopoli) – Spain, Teruel, Cosa, 10.vii.1986, gent. CG 2198.

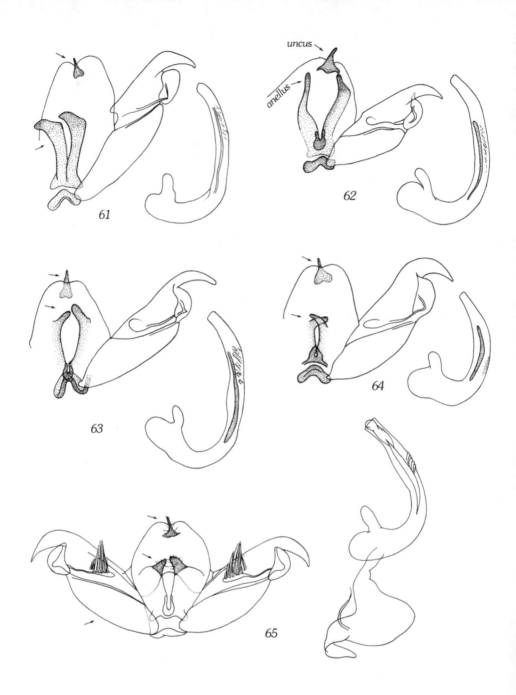

MALE GENITALIA

61. *Stenoptilia elkefi* Arenberger – Turkey, Van, Gölü, 1800 m, 25.vi.1965 (Noack), gent. CG 6252 (Löbb. Mus.).
62. *S. lucasi* Arenberger – Turkey, Gümüshane, Spikör Geçidi, 2390-2500 m, 26.vii.1989 (J.A.W. Lucas), gent. Ar 3626 (Lucas).
63. *S. annadactyla* Sutter – Switzerland, Graubünden, Löbbia, 28.vi.1984, gent. CG 2419.
64. *S. pelidnodactyla* (Stein) – Spain, Teruel, Sierra Alta, 1700 m, 25.vi.1987, gent. CG 2406.
65. *S. reisseri* Rebel – Spain, Avila, Sierra de Gredos, Gargante de la Pozas, 1900 m, 14.vii.1934 (Reisser), gent. Arenberger 3579 (LNK).

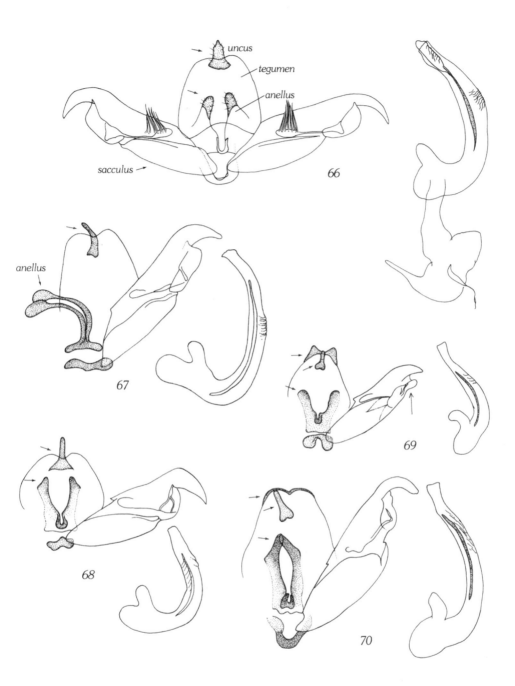

MALE GENITALIA

66. *Stenoptilia hahni* Arenberger – Spain, Granada, Sierra Nevada, Veleta Road, 1700 m, 9.vii.1971 (E. Arenberger), gent. Arenberger 2836 (E. Arenberger).

67. *S. millieridactyla* (Bruand) – Ireland, Dublin, 12.vi.1933, gent. BM 4804 (BMNH).

68. *S. islandicus* (Staudinger) – Iceland, gent. Gibeaux 2675 (ZMUC).

69. *S. parnasia* Arenberger – Greece, Peloponnesos, Zachlorau near Kalavryta, 600 m, 1-14.vii.1959 (Noach), gent. CG 2335.

70. *S. coprodactylus* (Stainton) – Turkey, Erzincan, Sakaltutan Geçidi, 2100-2200 m, 16.vii.1987 (W.O. de Prins), gent. CG 1922 (W.O. de Prins).

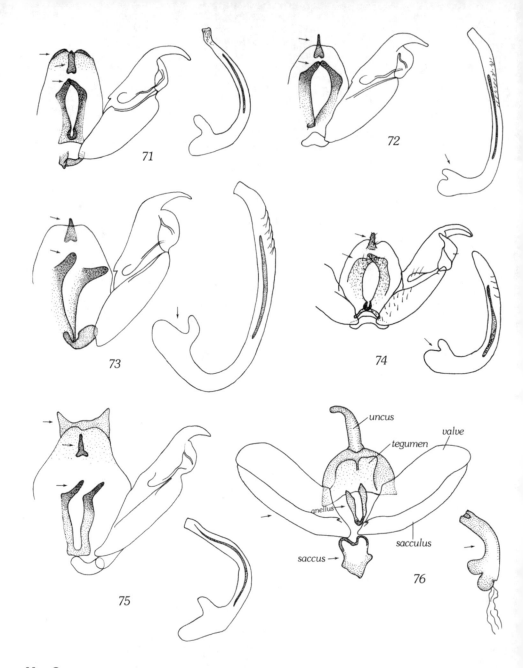

MALE GENITALIA

71. *Stenoptilia lutescens* (Herrich-Schäffer) – France, Alpes de Hte Provence, Col de Vars, 2200 m, 20-26.vii.1987, gent. CG 2405.
72. *S. nepetellae* Bigot & Picard – France, Vaucluse, Mont Ventoux, 22.vi.1985 (H.W. van der Wolf), gent. CG 2177.
73. *S. stigmatodactylus* (Zeller) – Italy, Mont Aurunci, 5 km N of Itri Latina, 4-11.viii.1972 (R. Johansson), gent. CG 2164.
74. *S. stigmatoides* Sutter & Skyva – After Sutter & Skyva, 1992.
75. *S. zophodactylus* (Duponchel) – Spain, Navarra, Sada, 23.vii.1985, gent. CG 2416.
76. *Buszkoiana capnodactylus* (Zeller) – Netherlands, Limburg, Schinveld, 2.vii.1974 (A. Schreurs), gent. CG 6204 (ITZ).

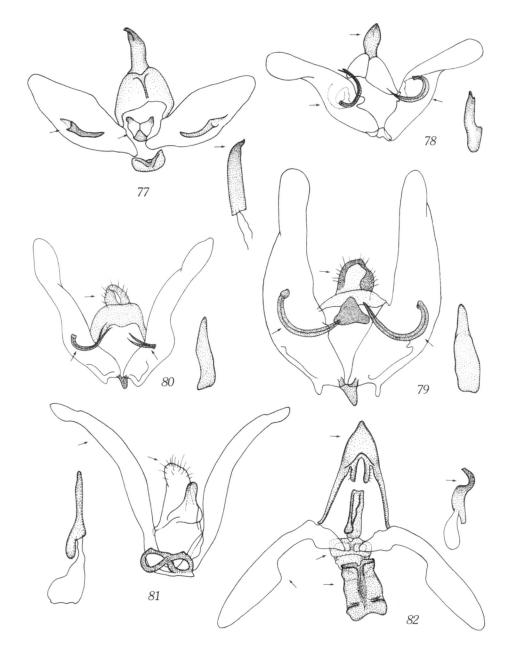

MALE GENITALIA

77. *Cnaemidophorus rhododactyla* ([Denis & Schiffermüller]) – Spain, Huesca, Lanave, 5.vii.1982, gent. CG 2003.

78. *Marasmarcha lunaedactyla* (Haworth) – France, Alpes Mar., Col de Couillole, 1600 m, 19.vii.1987, gent. CG 2236.

79. *M. fauna* (Millière) – France, Drôme, La Penne sur Ouvèze, 15-19.vii.1986 (H.W. van der Wolf), gent. CG 1842.

80. *M. oxydactylus* (Staudinger) – France, Drôme, Col de Soubeyrand, 995 m, 2.viii.1986 (H.W. van der Wolf), gent. CG 1841.

81. *Geina didactyla* (Linnaeus) – Spain, Teruel, Noguera, 1500 m, 11.vii.1986, gent. CG 2187.

82. *Procapperia maculatus* (Constant) – France, Hte Alpes, Reotier, 1000 m, 21.vi.1990 (A. Cox), gent. CG 6206 (Cox).

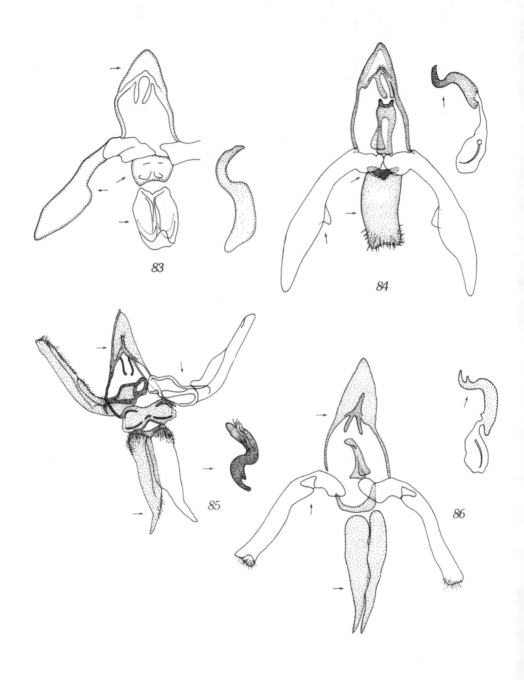

MALE GENITALIA

83. *Procapperia croatica* Adamczewski – After Adamczewski, 1951.
84. *Paracapperia anatolicus* (Caradja) – Turkey, E Anatolia, Gürün, 17.vi-3.vii.1976 (Pinker), gent. CG 2314.
85. *Capperia britanniodactylus* (Gregson) – Netherlands, Noord Holland, Santpoort, 18.vii.1984 (van Driel & E.J. van Nieukerken), gent. CG 1806.
86. *C. celeusi* (Frey) – France, Hte Alpes, near La Grave, 1400 m, 13.vii.1987, gent. CG 2229.

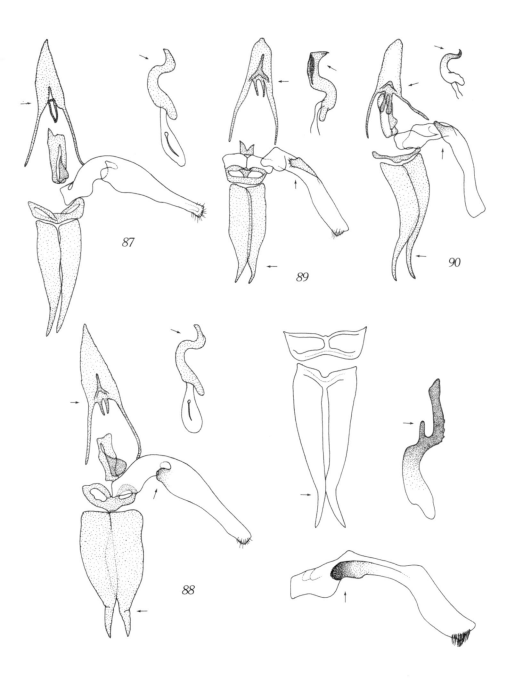

Male Genitalia

87. *Capparia trichodactyla* ([Denis & Schiffermüller]) – France, Isère, Col d'Ornon, 1370 m, 30.vii.1988, gent. CG 2252.
88. *C. fusca* (Hofmann) – Switzerland, Vaud, Glion, 900 m, 1-7.viii.1984 (A. Cox), gent. CG 2241.
89. *C. bonneaui* Bigot – Holotype: Spain, Teruel, Valdevecar, 1170 m, 20.vi.1986 (P. Bonneau), gent. Bigot 1014 (Bigot).
90. *C. hellenica* Adamczewski – France, Var, Taures, 420 m, e.l., 22.viii.1987 (J. Nel), gent. CG 2230.
91. *C. loranus* (Fuchs) – Bohemia, Radotin, 2.ix.1981 (Skyva), gent. CG 6166.

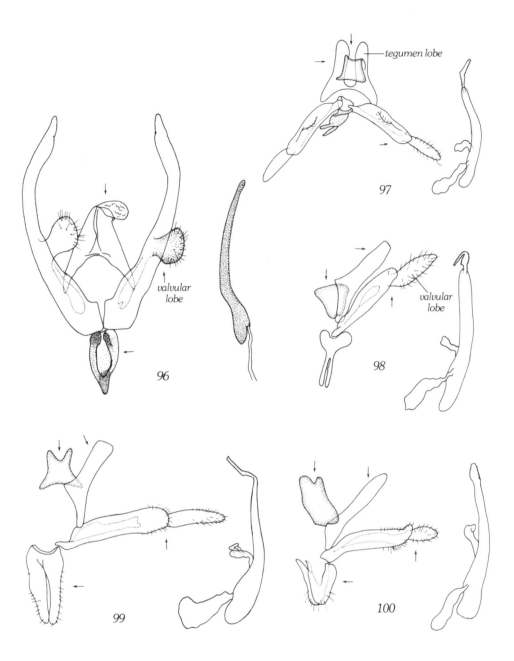

MALE GENITALIA

96. *Buckleria paludum* (Zeller) – Netherlands, Limburg, Brunssumer Heide, 18.viii.1983 (G. Langohr), gent. CG 2311.

97. *Oxyptilus pilosellae* (Zeller) – France, Isère, Freney d'Oisans, 25.vii-8.viii.1988, gent. CG 2301.

98. *O. chrysodactyla* ([Denis & Schiffermüller]) – Belgium, Brabant, Laken, 110 m, 26.vi.1969 (F. Coenen), gent. CG 2091.

99. *O. ericetorum* (Stainton) – France, Var, Ste Baume, 600 m, 18.vii.1988 (L. de Ridder), gent. CG 2307.

100. *O. parvidactyla* (Haworth) – France, Isère, Freney d'Oisans, 1500 m, 6-28.vii.1987, gent. CG 2228.

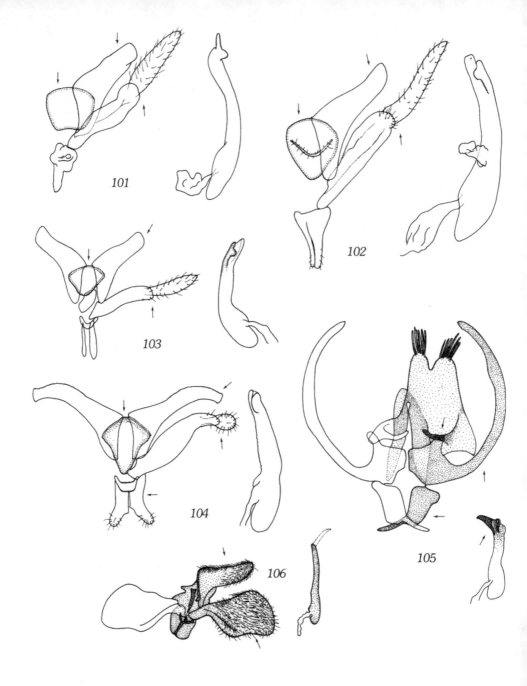

MALE GENITALIA

101. *Crombrugghia distans* (Zeller) – Austria, Nieder Österreich, Neu Aigen, 24.v.1986, gent. CG 2305.
102. *C. tristis* (Zeller) – Germania Sept, no date (Staudinger), gent. CG 3125 (NNM).
103. *C. kollari* (Stainton) – (Austria), Großglockner, no date (Mann), gent. CG 3122 (NNM).
104. *C. laetus* (Zeller) – France, Hte Alpes, Les Laus N of Col d'Izoard, 1800 m, 22.vii.1984 (J. Nel), gent. CG 2443.
105. *Stangeia siceliota* (Zeller) – Spain, Huelva, Calañas, 15.v.1981, gent. CG 2309.
106. *Megalorhipida leucodactylus* (Fabricius) – Côte d'Ivoire, Bouaflé, Bouaflé, 25.viii.1983 (R.T.A. Schouten), gent. CG 2168.

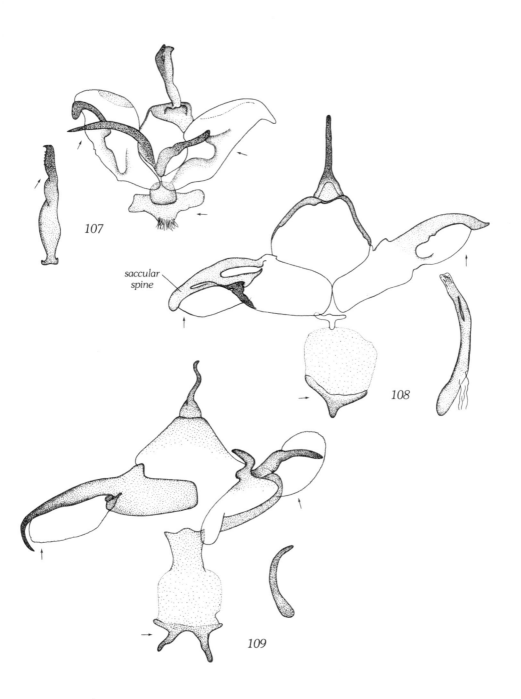

MALE GENITALIA

107. *Puerphorus olbiadactylus* (Millière) – Spain, Sevilla, Ronquillo, 15.v.1981, gent. CG 2060.
108. *Gypsochares baptodactylus* (Zeller) – Italy, Sardinia, Aritzo, 30.vii-4.viii.1981 (Kaltenbach), gent. CG 2359.
109. *G. bigoti* Gibeaux & Nel – Spain, Murcia, Murcia, 25.iv.1981, gent. CG 2357.

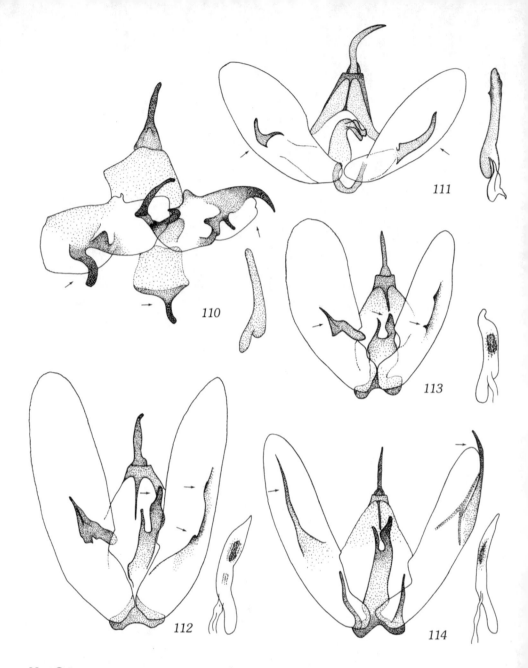

MALE GENITALIA

110. *Gypsochares nielswolffi* Gielis & Arenberger – Holotype: Madeira, Serra d'Agua, Station Salazar, 6-7.ix.1973 (Lomholdt & Wolff), gent. NLW 4201 (ZMUC).

111. *Pselnophorus heterodactyla* (Müller) – France, Vaucluse, Mont Ventoux, 1450 m, 5.vii.1989, gent. CG 2355.

112. *Hellinsia inulae* (Zeller) – Spain, Malaga, San Pedro de Alcantara, 18.iii.1986 (C. van Achterberg), gent. CG 2390.

113. *H. carphodactyla* (Hübner) – Netherlands, Limburg, Eygelshoven, 12.ix.1987 (A. Schreurs), gent. CG 1925 (H.W. van der Wolf).

114. *H. chrysocomae* (Ragonot) – France, Bouches-du-Rhône, La Ciotat, 380 m, 15.ix.1985 (J. Nel), gent. CG 2387.

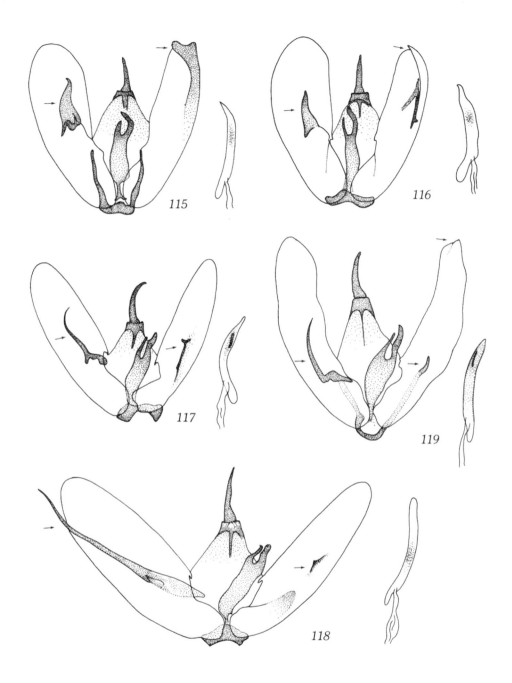

MALE GENITALIA

115. *Hellinsia osteodactylus* (Zeller) – France, Hte Rhine, Uftholz, 15.vii.1983 (W.O. de Prins), gent. CG 1774 (W.O. de Prins).
116. *H. pectodactylus* (Staudinger) – Spain, Valencia, Serra, 17.iv.1978, gent. CG 2428.
117. *H. distinctus* (Herrich-Schäffer) – Slovakia, Muran, 1000 m, 16-17.vii.1990 (H.W. van der Wolf).
118. *H. didactylites* (Ström) – Netherlands, Gelderland, Assel, 11.vi.1984, gent. CG 2445.
119. *H. tephradactyla* (Hübner) – Slovakia, Muran, 1000 m, 16-17.vii.1990 (H.W. van der Wolf), gent. CG 2429.

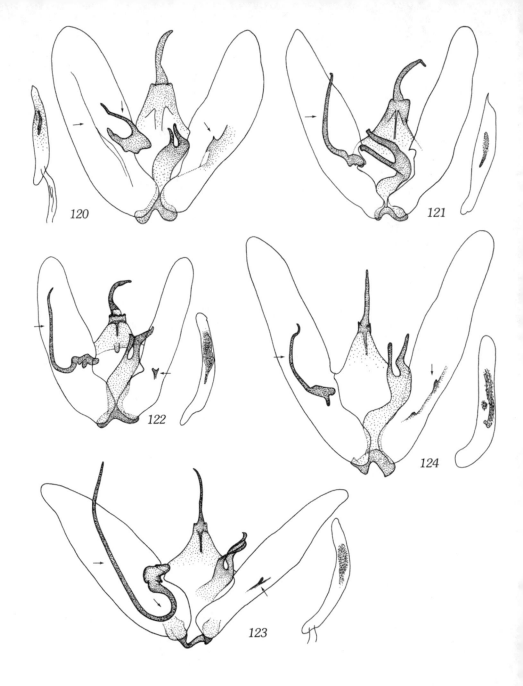

MALE GENITALIA

120. *Hellinsia lienigianus* (Zeller) – Belgium, Brabant, Rixensant, Grand Bruyère, 110 m, 10.vi.1981 (F. Coenen), gent. CG 2446.
121. *Oidaematophorus lithodactyla* (Treitschke) – Spain, Lerida, San Lorenzo de Morunys, 31.vii.1975, gent. CG 964.
122. *O. rogenhoferi* (Mann) – Norway, Indre Sør-Trøndelag, Kongsvoll, 900-1100 m, 20-28.vii.1983 (Karsholt & Michelsen), gent. CG 2376.
123. *O. constanti* (Ragonot) – France, Hte Alpes, Guillestre, 1050 m, e.l., 19.vii.1985 (J. Nel), gent. CG 2365.
124. *O. giganteus* (Mann) – France, Var, Col de Fourches, Maures, 17.v.1985 (J. Nel), gent. CG 2366.

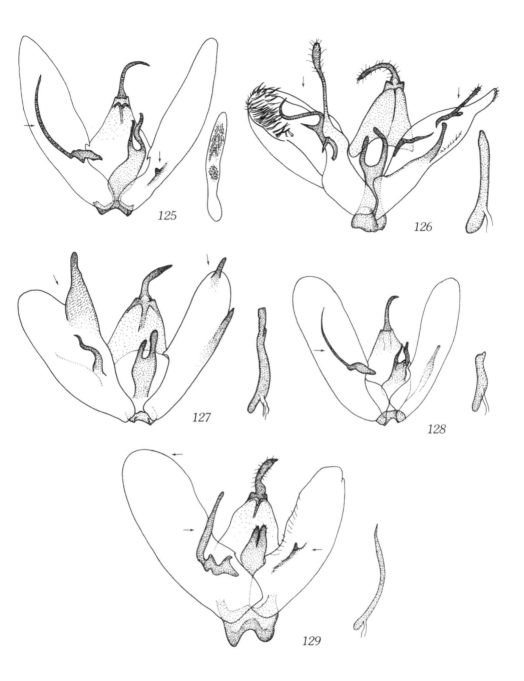

MALE GENITALIA

125. *Oidaematophorus vafradactylus* Svensson – Sweden, Gotland, Silte, 25.vii.1966 (R. Johansson), gent. CG 4200 (ZMUC).

126. *Emmelina monodactyla* (Linnaeus) – Greece, Lithochoron, 18.vii.1987 (H.W. van der Wolf), gent. CG 1890 (H.W. van der Wolf).

127. *E. argoteles* (Meyrick) – Germany, Hessen, Torfgrube S Pfungstadt, 16.viii.1964 (Groß), gent. CG 2370.

128. *Adaina microdactyla* (Hübner) – Spain, Huesca, Torla, 1100 m, 3.viii.1991 (H.W. van der Wolf), gent. CG 2517.

129. *Calyciphora punctinervis* (Constant) – Spain, Granada, Pto de Mora, 21.ix.1979, gent. CG 2410.

173

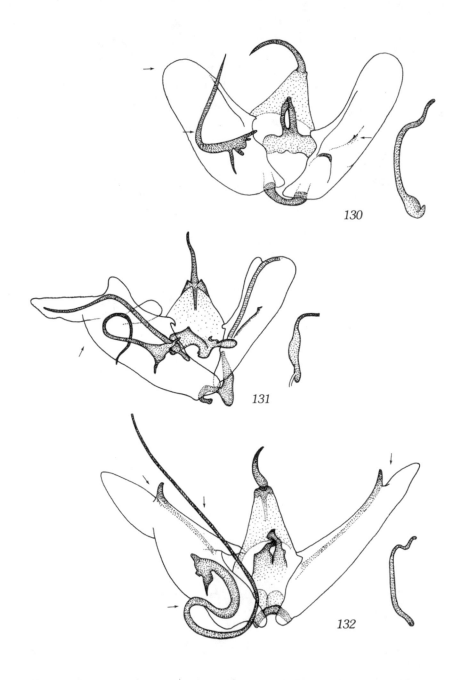

MALE GENITALIA

130. *Calyciphora homoiodactyla* (Kasy) – France, Hte Alpes, Prunière, 800 m, 1-15.vii.1985 (A. Cox), gent. CG 1900 (A. Cox).
131. *C. adamas* (Constant) – France, Ardèche, St. Martin, 10.ix.1985 (A. Cox), gent. 2345.
132. *C. acarnella* (Walsingham) – France, Corsica, Corté, 9.vi.1948, gent. BM 5001 (BMNH).

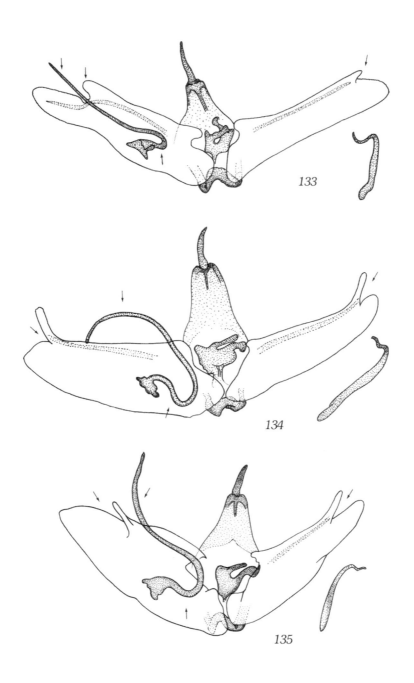

MALE GENITALIA

133. *Calyciphora albodactylus* (Fabricius) – Austria, Styria, Graz, 14.viii.1907, gent. CG 6266 (ITZ).
134. *C. xanthodactyla* (Treitschke) – Gent. Mus. Vind. 10640 (Mus. Vind.).
135. *C. nephelodactyla* (Eversmann) – Spain, 8 km SE of Laguna Dalga, 42°18' N 5°42' W, 1900 m, 2.viii.1988 (R.T.A. Schouten), gent. CG 2342.

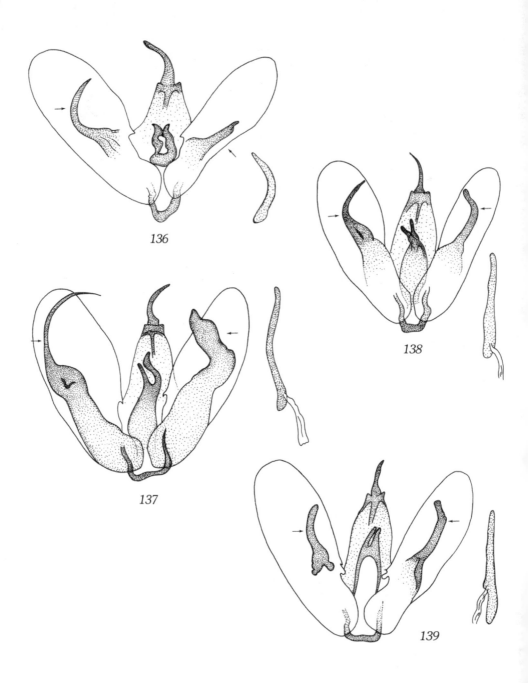

MALE GENITALIA

136. *Porrittia galactodactyla* ([Denis & Schiffermüller]) – Netherlands, Noord Holland, Overveen, 18-30.vi.1988, gent. CG 2350.
137. *Merrifieldia leucodactyla* ([Denis & Schiffermüller]) – Spain, Cuenca, Beteta, 2.vi.1982, gent. CG 2450.
138. *M. tridactyla* (Linnaeus) – Spain, Granada, Sierra Nevada, Veleta Road, 2000 m, 23-29.vii.1986, gent. CG 1898.
139. *M. baliodactylus* (Zeller) – Germany, Bayern, Kaiserwacht, 9.vii.1084, gent. CG 2338.

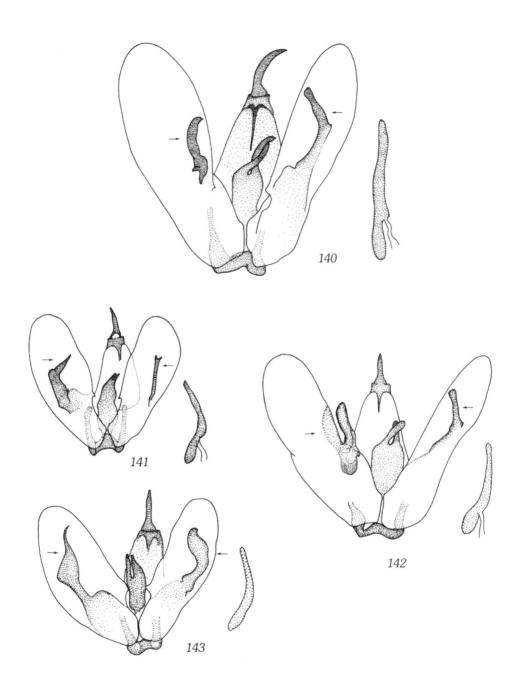

MALE GENITALIA

140. *Merrifieldia malacodactylus* (Zeller) – Tunisia, Ain Draham, 5-18.v.1988 (Zool. Mus. Exp. Copenhagen), gent. CG 4202 (ZMUC).
141. *M. semiodactylus* (Mann) – France, Corsica, Ajaccio, St. Antone, 80 m, 2.v.1988 (J. Nel), gent. CG 2409.
142. *M. hedemanni* (Rebel) – Canary Islands, Tenerife, Guimar, 26.iv.1907, gent. BM 15417 (BMNH).
143. *M. chordodactylus* (Staudinger) – Canary Islands, Tenerife, Sta. Cruz, 23.xii.1906, gent. BM 15416 (BMNH).

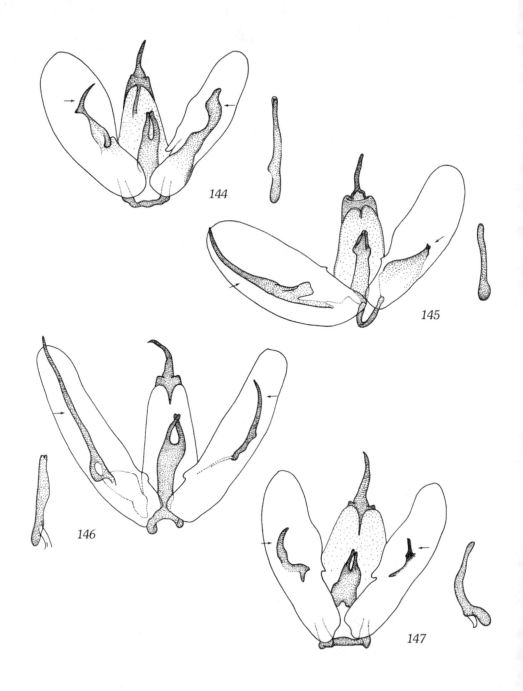

MALE GENITALIA

144. *Merrifieldia bystropogonis* (Walsingham) – Canary Islands, Palma, 9.iii.1972 (Stamm), gent. CG 2386.
145. *Wheeleria phlomidis* (Staudinger) – Turkey, Anatolia, Aksehir, 1000 m, 2.vii.1963 (Noach), gent. CG 2418.
146. *W. raphiodactyla* (Rebel) – Spain, Granada, Sierra Nevada, Veleta Road, 2000 m, 12.vii.1971 (E. Arenberger), gent. CG 2380.
147. *W. spilodactylus* (Curtis) – Great Britain, Wales, Gwynned, 5.vii.1984 (M. Hull), gent. CG 2088.

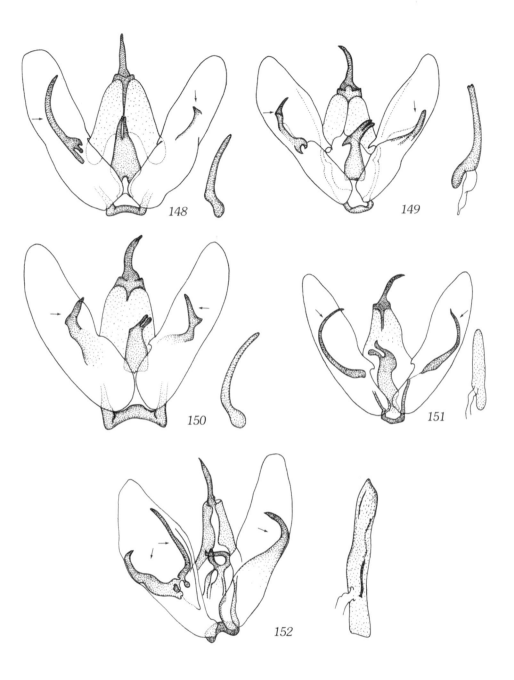

MALE GENITALIA

148. *Wheeleria obsoletus* (Zeller) – Syria, Shar Devesy, Haleb, 1893, gent. BM 15404 (BMNH).
149. *W. lyrae* (Arenberger) – Greece, Lakonia, Waterfall near Nomia-Lyra, 17.v.1979 (E. Arenberger), gent. Arenberger 1961 (MNMB).
150. *W. ivae* (Kasy) – Syria, Shar Devesy, Haleb, 1893, gent. BM 16052 (BMNH).
151. *Pterophorus pentadactyla* (Linnaeus) – France, 30 km N of Troyes, 82 m, 12.viii.1988 (R.T.A. Schouten), gent. CG 2337.
152. *P. ischnodactyla* (Treitschke) – South Africa, Bloemfontein, 30.i.1951 (van Ee), gent. CG 1695 (ITZ).

179

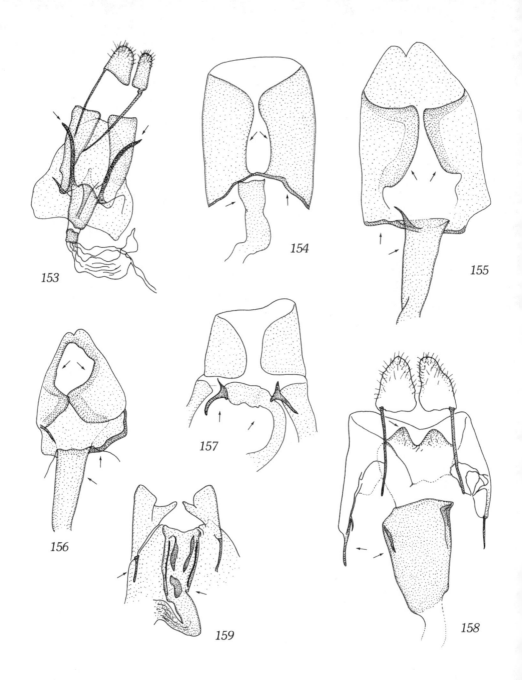

FEMALE GENITALIA

153. *Agdistis tamaricis* (Zeller) – Spain, Malaga, Marbella, 5.v.1981, gent. CG 1431.
154. *A. intermedia* Caradja – Hungary, Hortobagy Nat. Park, E of Nagyvan, 10-11.vii.1983 (M. & E. Arenberger), gent. CG 2437.
155. *A. bennetii* (Curtis) – Spain, Huelva, El Rompido, 13.v.1979, gent. CG 2379.
156. *A. meridionalis* (Zeller) – France, Cannes, 9.vi.1884, gent. CG 3022 (NNM).
157. *A. salsolae* Walsingham – Canary Islands, Fuerteventura, Corralejo, 3-8.ii.1985 (A. Cox), gent. CG 2210.
158. *A. delicatulella* Chrétien – Malta, Zebbuq, 6.xi.1983 (P. Sammut), gent. CG 1545.
159. *A. neglecta* Arenberger – France, Alpes Mar., no date (Millière), gent. CG 3014 (NNM).

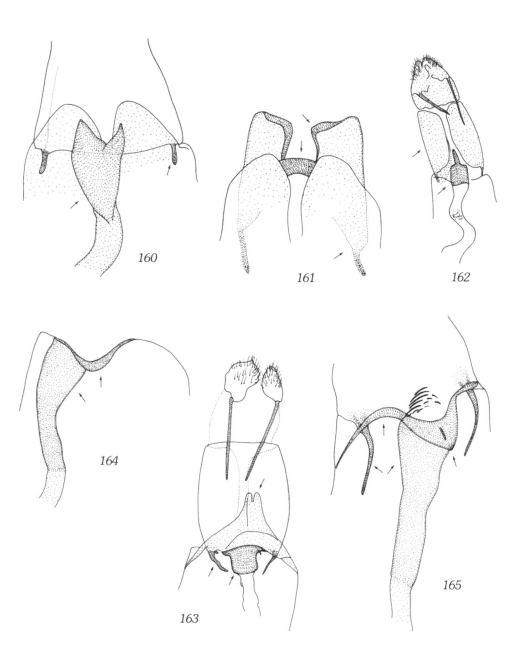

FEMALE GENITALIA

160. *Agdistis protai* Arenberger – Italy, Sardinia, gent. Arenberger 468 (Arenberger).
161. *A. adactyla* (Hübner) – Russia, Puvocias, 13.vii.1979 (P. Ivinskis), gent. CG 2213.
162. *A. heydeni* (Zeller) – Spain, Huelva, El Rompido, 13.v.1981, gent. CG 1433.
163. *A. satanas* Millière – Malta, Rabat, 26.viii.1983 (P. Sammut), gent. CG 1544.
164. *A. frankeniae* (Zeller) – Spain, Huelva, El Rompido, 15.v.1981, gent. CG 2201.
165. *A. gittia* Arenberger – Spain, Granada, Baza, 110 km NE Granada, 9-10.v.1977 (M. & W. Glaser), gent. Arenberger 2682 (Arenberger).

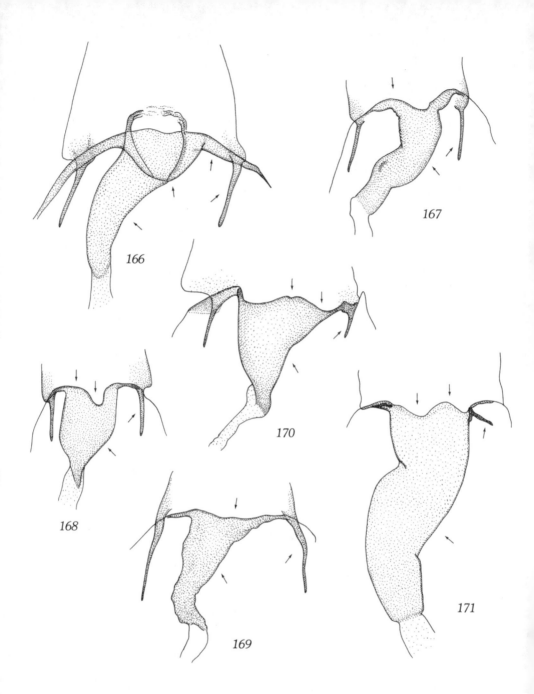

166. *Agdistis espunae* Arenberger – Spain, Murcia, Alhama de Murcia, 11.vi.1974 (M. & W. Glaser), gent. Arenberger 1527 (Arenberger).

167. *A. glaseri* Arenberger – Spain, Murcia, Aguilas, 22.iv.1981, gent. CG 2200.

168. *A. bigoti* Arenberger – Greece, Crete, Matala, 25.v.1965 (F. Tondeur), gent. Arenberger 445 (Arenberger).

169. *A. symmetrica* Amsel – Gent. Arenberger 2018 (Mus. Vind.).

170. *A. manicata* Staudinger – Spain, Cadiz, Chiclana, 29.ix.1983, gent. CG 2384.

171. *A. paralia* (Zeller) – France, Camargue, Valas, 17.viii.1961 (L. Bigot), gent. CG 2364.

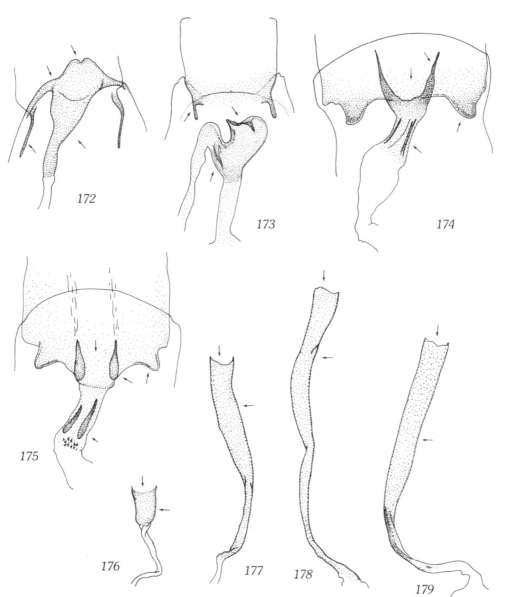

FEMALE GENITALIA

172. *Agdistis bifurcatus* Agenjo – Spain, Huelva, Huelva, 2.x.1958 (von Buddenbrock), gent. CG 5107 (Löbb. Mus.).

173. *A. pseudocanariensis* Arenberger – Canary Islands, Tenerife, El Medano, 7.xi.1970 (Pinker), gent. Arenberger 1379 (Arenberger).

174. *A. hartigi* Arenberger – Spain, Almeria, 6 km SW of Tabernas, Mini Hollywood, 400 m, 10.x.1993 (H.W. van der Wolf), gent. CG 2514.

175. *A. betica* Arenberger – Spain, Granada, Baza, 110 km NE of Granada, 2.vi.1975 (M. & W. Glaser), gent. Arenberger 1554 (Arenberger).

176. *Platyptilia tesseradactyla* (Linnaeus) – Canada, British Columbia, Clinton, 12.vi.1938 (J.K. Jakob), gent. CG 2324.

177. *P. farfarellus* Zeller – Italy, Garda, Voltino di Tremosin, 30.v.1983 (J. Neyts), gent. CG 2030.

178. *P. nemoralis* Zeller – Belgium, Liège, Mont Rigi, 670 m, 25.vii-3.viii.1964 (Ent. Exc. Zool. Mus.), gent. CG 2441.

179. *P. gonodactyla* ([Denis & Schiffermüller]) – Netherlands, Zuid Holland, Oegstgeester Polder, 21.v.1988 (R.T.A. Schouten), gent. CG 2325.

183

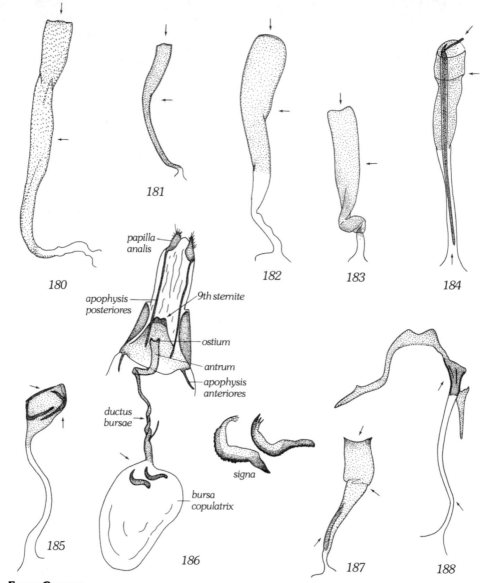

FEMALE GENITALIA

180. *Platyptilia calodactyla* ([Denis & Schiffermüller]) – Netherlands, Zuid Holland, Vlaardingen, 12.vii.1976, gent. CG 1002.

181. *P. iberica* Rebel – Spain, Granada, Sierra Nevada, Pto de Lobo, 2100 m, 21.vii.1937 (Reisser), gent. CG 1752 (Mus. Vind.).

182. *P. isodactylus* (Zeller) – Netherlands, Friesland, Terschelling, Nieuw Formenum, 2.ix.1984 (JP & LB), gent. CG 2175.

183. *Gillmeria miantodactylus* (Zeller) – East Europe, no date, gent. BM 14168 (BMNH).

184. *G. pallidactyla* (Haworth) – Netherlands, Limburg, St Joost, 25.vii.1978, gent. CG 1187.

185. *G. tetradactyla* (Linnaeus). Lectotype female of *P. borgmanni* Roessler. – (Germany, Wiesbaden), no date, gent. Mus. Wiesbaden 151 (Mus. Wiesbaden).

186. *Lantanophaga pusillidactylus* (Walker) – Morocco, Rabat, Jardin d'Essai, 20.vii.1953 (Ch. Rungs), gent. CG 1670 (MNHN).

187. *Stenoptilodes taprobanes* (Felder & Rogenhofer) – Tchad, Bededija, near Moundou, 400 m, 30.x.1970 (J.H. & M. Lourens), gent. CG 2001.

188. *Paraplatyptilia metzneri* (Zeller) – France, Hte Alpes, Col de Vars, 27.vii.1970 (E. Arenberger), gent. CG 2404.

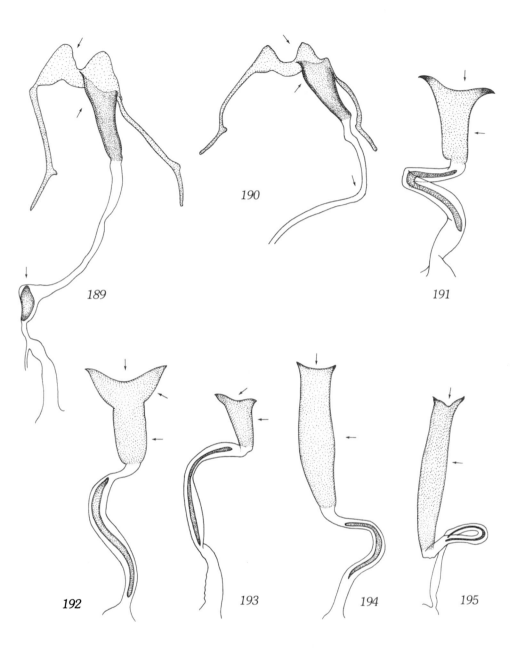

FEMALE GENITALIA

189. *Amblyptilia acanthadactyla* (Hübner) – Spain, Murcia, Mazarron, 26.iv.1981, gent. CG 2322.

190. *A. punctidactyla* (Haworth) – Austria, Turracher Höhe, 7.vii.1984, gent. CG 2440.

191. *Stenoptilia graphodactyla* (Treitschke) – Austria, Salzburg, Torrenerjoch, 1700 m, 20.vii.1920 (Kautz), gent. CG 1755 (Mus. Vind.).

192. *S. pneumonanthes* (Büttner) – Netherlands, Friesland, Appelscha, 17.viii.1980, gent. CG 1518.

193. *S. gratiolae* Gibeaux & Nel – Pommern, no date, gent. Bigot 18 (MNHN).

194. *S. pterodactyla* (Linnaeus) – France, Clermont Ferrand, 14.vii.1975, gent. CG 973.

195. *S. mannii* (Zeller). Lectotype female of *S. megalochra* Meyrick. – Bulgaria, Rila Kloster, viii.(19)11, gent. BM 15426 (BMNH).

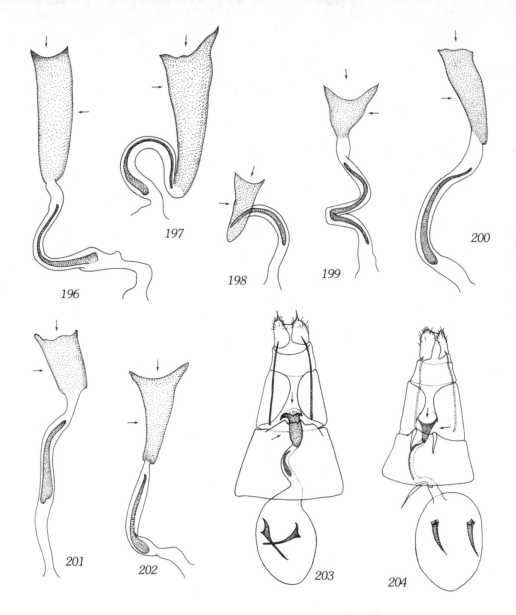

FEMALE GENITALIA

196. *Stenoptilia veronicae* Karvonen – Finland, Lapponia inarensis, Ingri, 18.vii.1985 (G. Langohr), gent. CG 2332.

197. *S. bipunctidactyla* (Scopoli) – Madeira, Funchal, Lido, 4-14.ix.1973 (Lomholdt & Wolff), gent. CG 4203 (ZMUC).

198. *S. aridus* Zeller – Spain, Cuenca, Tejadillo, 15.vii.1986, gent. CG 2197.

199. *S. elkefi* Arenberger – Greece, Chalkidike, Amaia, 900 m, 6.viii.1966 (Lukasch), gent. CG 2090.

200. *S. lucasi* Arenberger – Turkey, Gümüshane, Spika Geçidi, 2350-2500 m, 26.vii.1989 (J.A.W. Lucas), gent. CG 6244 (J.A.W. Lucas).

201. *S. annadactyla* Sutter – Italy, Toscana, Lucotena, 12.viii.1982 (J.H. Kuchlein), gent. CG 2007.

202. *S. pelidnodactyla* (Stein) – Germany, Bayern, Graswang, 10.vii.1984, gent. CG 2037.

203. *S. reisseri* Rebel. Type female. – Spain, Avila, Sierra de Gredos, Garganta de la Pozas, 1900 m, 11.vii.1934 (Reisser), gent. Jächk 4727 (Mus. Vind.). After Arenberger.

204. *S. hahni* Arenberger – Spain, Granada, Sierra Nevada, Veleta Road, 2200 m, 8.vii.1979 (Hahn). After Arenberger, 1990.

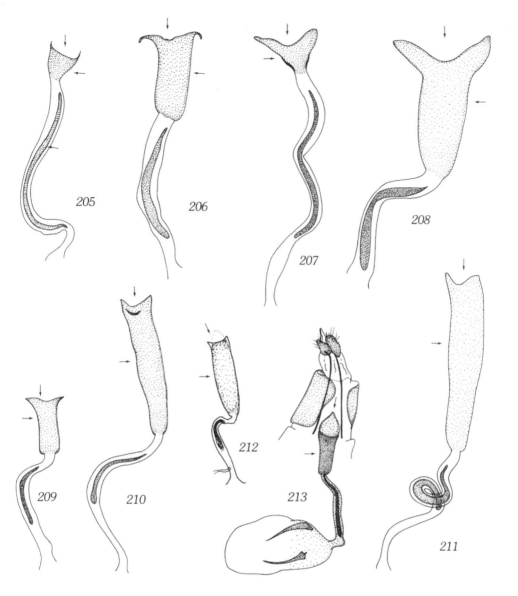

FEMALE GENITALIA

205. *Stenoptilia millieridactyla* (Bruand) – Spain, Huesca, Sabiñanigo, 13.vi.1982 (F. Coenen), gent. CG 1716 (MNHN).

206. *S. islandicus* (Staudinger) – Iceland, gent. Gibeaux 2676 (ZMUC).

207. *S. parnasia* Arenberger – Greece, Lakonia, Taygetos Mtns, S of Sparta, Paleopanagia, 1600 m, 10.viii.1985 (M. & E. Arenberger), gent. Arenberger 773 (Arenberger).

208. *S. coprodactylus* (Stainton) – France, Savoie, 8 km N of Col du Galibier, 2000 m, 27.vii.1988, gent. CG 2408.

209. *S. lutescens* (Herrich-Schäffer) – France, Alpes de Hte Provence, Col de Vars, 1650-1900 m, 20-26.vii.1987, gent. CG 2331.

210. *S. nepetellae* Bigot & Picard – France, Vaucluse, Mont Ventoux, 21.vi.1985 (H.W. van der Wolf), gent. CG 2330.

211. *S. stigmatodactylus* (Zeller) – Italy, Mont Aurunci, 5 km N of Itri Latina, 4-11.viii.1972 (R. Johansson), gent. CG 2165.

212. *S. stigmatoides* Sutter & Skyva – After Sutter & Skyva, 1992.

213. *S. zophodactylus* (Duponchel) – Belgium, Antwerpen, Niel, 9.vi.1988 (W.O. de Prins), gent. CG 1967 (W.O. de Prins).

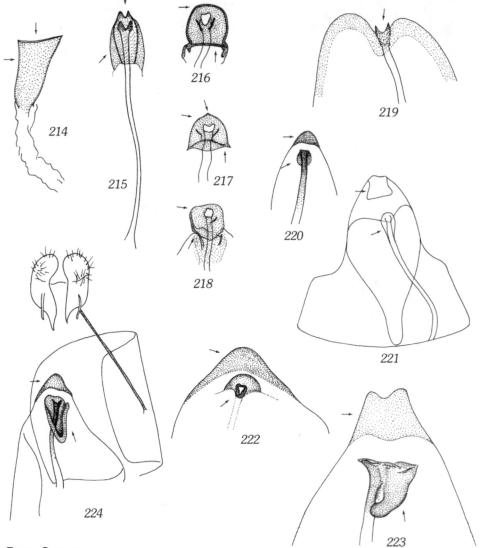

FEMALE GENITALIA

214. *Buszkoiana capnodactylus* (Zeller) – Netherlands, Limburg, Schinnen, 13.vii.1973 (G. Langohr), gent. CG 6203 (ITZ).

215. *Cnaemidophorus rhododactyla* ([Denis & Schiffermüller]) – France, Drôme, La Penne sur Ouvèze, 10-23.vii.1988, gent. CG 2320.

216. *Marasmarcha lunaedactyla* (Haworth) – France, Vosges, Coussey, 29.vii.1984 (H.W. van der Wolf), gent. CG 1659 (H.W. van edr Wolf).

217. *M. fauna* (Millière) – France, Drôme, La Penne sur Ouvèze, 20.vii.1984 (H.W. van der Wolf), gent. CG 2058.

218. *M. oxydactylus* (Staudinger) – Andorra, El Serrat, 1700 m, 16.vii.1981 (R.T.A. Schouten), gent. CG 2174.

219. *Geina didactyla* (Linnaeus) – France, Hte Alpes, Risoul, 2200 m, 6-7.viii.1984 (A. Cox), gent. CG 6118 (A. Cox).

220. *Procapperia maculatus* (Constant) – France, Savoie, Valloire, 1600 m, 10-13.vii.1987, gent. CG 2313.

221. *P. croatica* Adamczewski – After Adamczewski, 1951.

222. *Paracapperia anatolicus* (Caradja) – Turkey, Konya, 15 km S of Karaman, 1200 m, 19.vii.1985 (W.O. de Prins), gent. CG 2167.

223. *Capperia britanniodactylus* (Gregson) – Netherlands, Limburg, Brunssumer Heide, 13.vii.1985 (A. Schreurs), gent. CG 2316.

224. *C. celeusi* (Frey) – France, Isère, Freney d'Oisans, 1000 m, 6-28.vii.1987, gent. CG 2226.

188

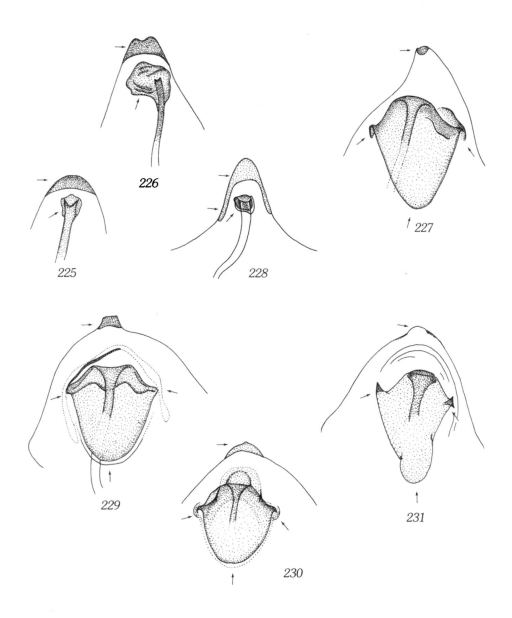

FEMALE GENITALIA

225. *Capperia trichodactyla* ([Denis & Schiffermüller]) – France, Savoie, 8 km N of Col du Galibier, 2000 m, 27.vii.1988, gent. CG 2439.

226. *C. fusca* (Hofmann) – (France), Moulin, e.l., 6.vi.1897, gent. BM 5403 (BMNH).

227. *C. bonneaui* Bigot – Spain, Teruel, Albarracin, Valdevecar, 1100 m, 30.vii.1991 (H.W. van der Wolf), gent. CG 2518.

228. *C. hellenica* Adamczewski – Spain, Teruel, Valdeltormo, 27.vi.1976, gent. CG 2318.

229. *C. loranus* (Fuchs) – Slovakia, Zadiel, 18-20.vii.1990 (H.W. van der Wolf), gent. CG 6200.

230. *C. marginellus* (Zeller) – (Italy), Syracuse, v (Zeller), gent. BM 5405 (BMNH).

231. *C. polonica* Adamczewski – France, Bouches-du-Rhône, Pomegues, 23.vi.1978 (Picard), gent. CG 2231.

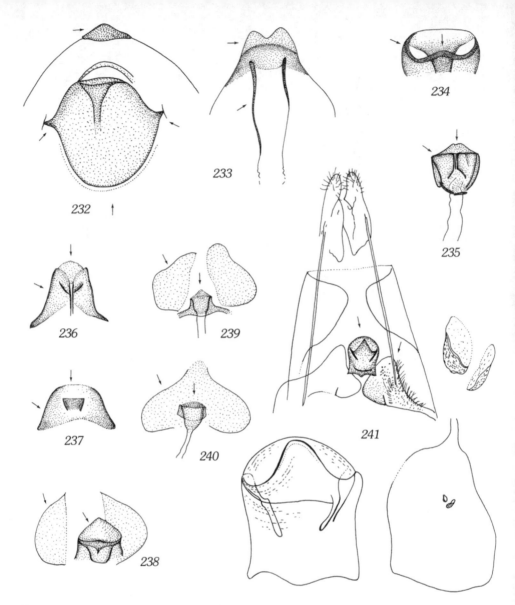

FEMALE GENITALIA

232. *Capperia maratonica* Adamczewski – Turkey, Kars, 7-10 km SE of Sarikanis, 2000 m, 20-24.vii.1990 (W.O. de Prins), gent. CG 2272.

233. *Buckleria paludum* (Zeller) – Netherlands, Limburg, Brunssumer Heide, 16.viii.1983 (G. Langohr), gent. CG 2310.

234. *Oxyptilus pilosellae* (Zeller) – Czechia, Moravia, Havraniky near Znojmo, 22.viii.1990 (H.W. van der Wolf), gent. CG 6198 (H.W. van der Wolf).

235. *O. chrysodactyla* ([Denis & Schiffermüller]) – Netherlands, Limburg, Wrakelberg, 9.vi.1983 (G. Langohr), gent. CG 2302.

236. *O. ericetorum* (Stainton) – France, Var, Ste Baume, 600 m, 5.viii.1986 (L. de Ridder), gent. CG 2304.

237. *O. parvidactyla* (Haworth) – Spain, Cuenca, Mota de Cuervo, 18.v.1981, gent. CG 2173.

238. *Crombrugghia distans* (Zeller) – France, Drôme, La Penne sur Ouvèze, 25-27.vii.1986 (H.W. van der Wolf), gent. CG 1840 (H.W. van der Wolf).

239. *C. tristis* (Zeller) – Germany, Friedland, no date, gent. BM 13171 (BMNH).

240. *C. kollari* (Stainton) – Austria, Großglockner, no date (Mann), gent. CG 3123 (NNM).

241. *C. laetus* (Zeller) – Spain, Gerona, Bagur, 25.v.1983 (A. Cox), gent. CG 1901.

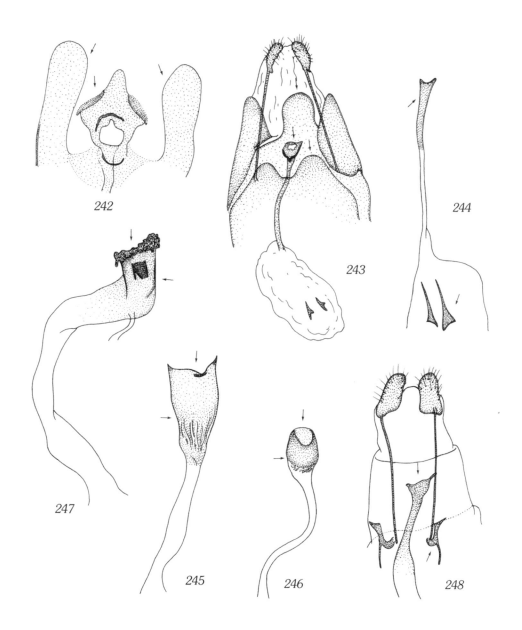

FEMALE GENITALIA

242. *Stangeia siceliota* (Zeller) – Switzerland, Vaud, Glion, 900 m, 15-30.vii.1984 (A. Cox), gent. CG 2208.

243. *Megalorhipida leucodactylus* (Fabricius) – Côte d'Ivoire, Bouaflé, Bouaflé, 11.viii.1983 (R.T.A. Schouten), gent. CG 2312.

244. *Puerphorus olbiadactylus* (Millière) – Spain, Huelva, Calañas, 15.v.1981, gent. CG 2356.

245. *Gypsochares baptodactylus* (Zeller) – Italy, Sardinia, Aritzo, 30.vii-4.viii.1981 (Kaltenbach), gent. CG 2360.

246. *G. bigoti* Gibeaux & Nel – Spain, Huelva, Calañas, 15.v.1981, gent. CG 2358.

247. *Pselnophorus heterodactyla* (Müller) – France, Vaucluse, Mont Ventoux, 1400 m, 18-24.vii.1988, gent. CG 2354.

248. *Hellinsia inulae* (Zeller) – Moravia, Marefy, 27.viii.1978 (Marek), gent. CG 2447.

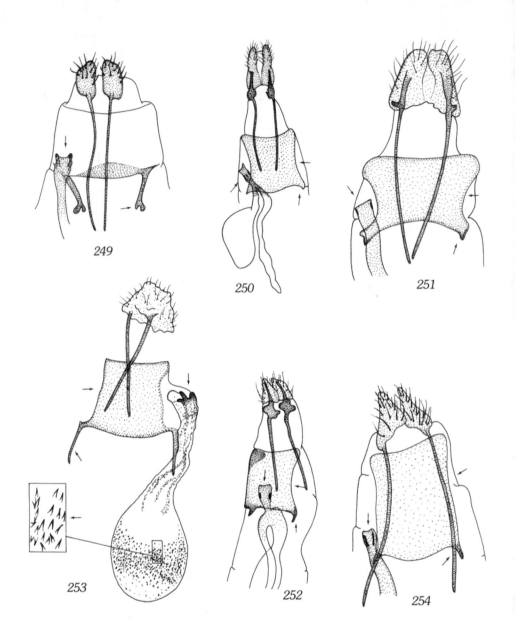

249

250

251

253

252

254

FEMALE GENITALIA

249. *Hellinsia carphodactyla* (Hübner) – Germany, Bayern, Fürstenfeldbruck, 16.v-12.vi.1986, gent. CG 2362.
250. *H. chrysocomae* (Ragonot). Lectotype, gent. Mus. Berlin 4587 (MNHN).
251. *H. osteodactylus* (Zeller) – Sweden, Hälsingland, Ljusdal Färila, 4.vii.1978 (Groß), gent. CG 5121 (Löbb. Mus.).
252. *H. pectodactylus* (Staudinger) – Germany, Hessen, Rheingau, Nollig near Lorch, 30.viii.1962 (Groß), gent. CG 5123 (Löbb. Mus.).
253. *H. distinctus* (Herrich-Schäffer), dorsal view – Netherlands, Gelderland, Imbosch, 31.vii.1975 (A. Cox), gent. CG not numbered (A. Cox).
254. *H. didactylites* (Ström) – Netherlands, Noord Brabant, Rijsbergen, De Pannenhoef, 14.vii.1973 (M. Hull), gent. Hull 765 (M. Hull).

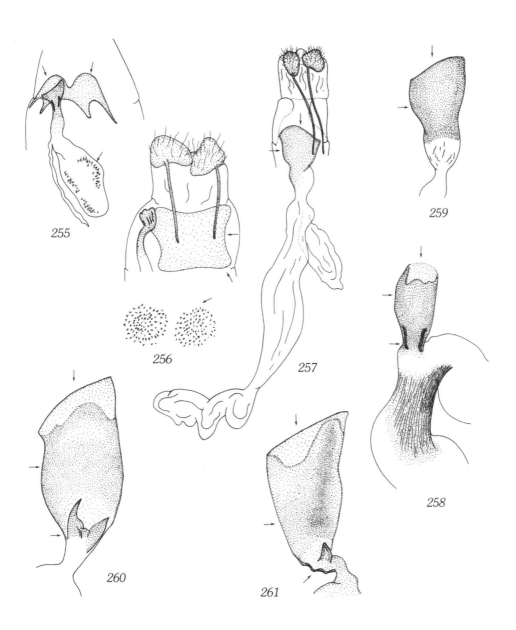

FEMALE GENITALIA

255. *Hellinsia tephradactyla* (Hübner) – France, Savoie, 8 km N of Col du Galibier, 2000 m, 27.vii.1988, gent. CG 2411.

256. *H. lienigianus* (Zeller) – France, Ardèche, Les Vans, 22-27.vi.1986 (D. Teunissen), gent. CG 2217.

257. *Oidaematophorus lithodactyla* (Treitschke) – France, Drôme, Col d'Ey, 720 m, 11.viii.1988, gent. CG 2244.

258. *O. rogenhoferi* (Mann) – France, Hte Alpes, Clarée Plampinet, 1500 m, 7.ix.1982 (C. Gibeaux), gent. CG 2363.

259. *O. constanti* (Ragonot) – France, Var, Plan d'Amps, 750 m, 2.vii.1985 (J. Nel), gent. CG 2417.

260. *O. giganteus* (Mann) – France, Var, Six-Fours, 16.v.1985 (J. Nel), gent. CG 2367.

261. *O. vafradactylus* Svensson – Sweden, Öland, Durstad, 7.viii.1984 (B.Å. Bengtsson), gent. CG 2368.

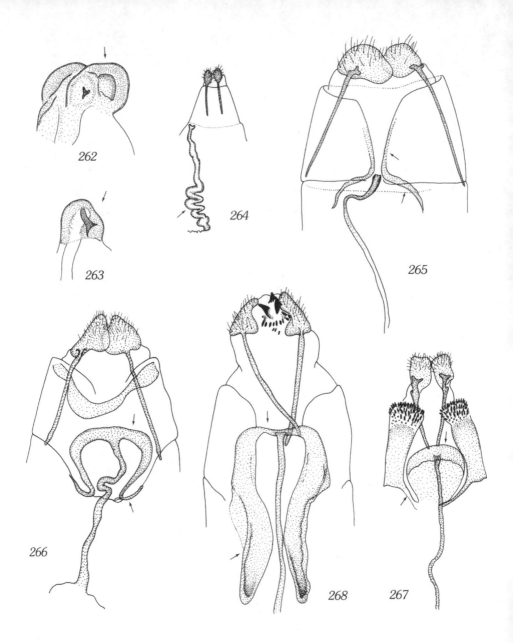

FEMALE GENITALIA

262. *Emmelina monodactyla* (Linnaeus) – U.S.A., California, Running Springs, 1500 m, 28.vii.1988 (H.W. van der Wolf), gent. CG 2258.

263. *E. argoteles* (Meyrick) – Japan, Honshu, Kawagana prefecture, Hayama, 1.vii.1985 (M. Takasu), gent. CG 6210 (ITZ).

264. *Adaina microdactyla* (Hübner) – Indonesia, Sulawesi, Toraja district, 1-5.vi.1986 (G. Gielis), gent. CG 2221.

265. *Calyciphora punctinervis* (Constant) – France, Bouches-du-Rhône, La Ciotat, 170 m, 4.ix.1988 (J. Nel), gent. CG 2349.

266. *C. homoiodactyla* (Kasy) – Turkey, Kirklareli, Ineci, 200 m, 15-16.vi.1988 (Wagener), gent. CG 3387.

267. *C. adamas* (Constant) – Type female: gent. Gibeaux 2376 (MNHN).

268. *C. acarnella* (Walsingham) – Italy, Sardinia, Aritzo, 30.vii-4.viii.1981 (Kaltenbach), gent. CG 2078.

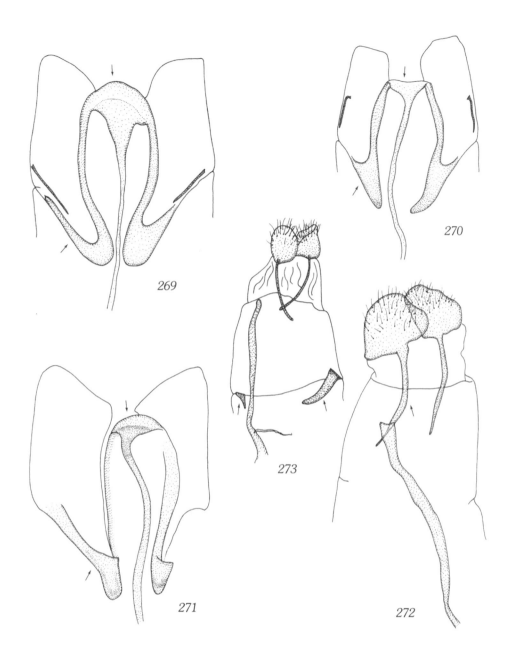

FEMALE GENITALIA

269. *Calyciphora albodactylus* (Fabricius) – Spain, Teruel, Cosa, 2-13.viii.1989, gent. CG 2341.

270. *C. xanthodactyla* (Treitschke) – No locality, gent. Mus. Vind. 10699 (Mus. Vind.).

271. *C. nephelodactyla* (Eversmann) – France, Alpes Mar., 10 km N of Uzelle, 8.viii.1987 (E.J. van Nieukerken), gent. CG 2343.

272. *Porrittia galactodactyla* ([Denis & Schiffermüller]) – Netherlands, Noord Holland, Overveen, 18-30.vi.1987, gent. CG 2351.

273. *Merrifieldia leucodactyla* ([Denis & Schiffermüller]) – Germany, Bayern, Graswang, 10.vii.1984, gent. CG 2413.

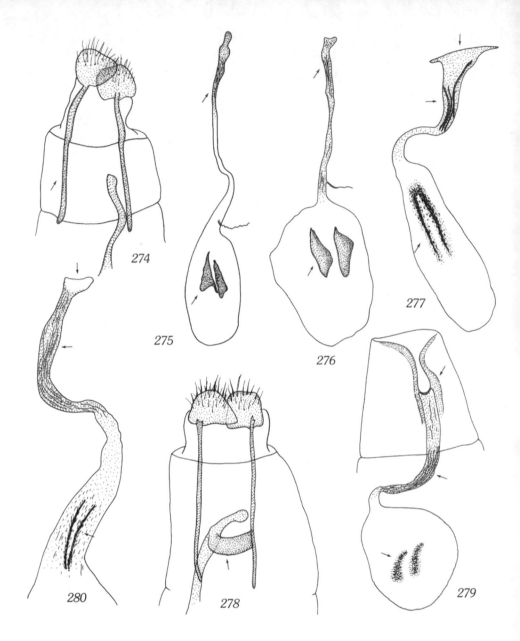

FEMALE GENITALIA

274. *Merrifieldia tridactyla* (Linnaeus) – Spain, Granada, Sierra Nevada, Veleta Road, 2000 m, 23-29.vii.1986, gent. CG 2383.

275. *M. baliodactylus* (Zeller) – Germany, Bayern, Kaiserwacht, 9.vii.1984, gent. CG 2339.

276. *M. malacodactylus* (Zeller) – "Yugoslavia", Dalmatia, Baska Voda, 18-28.ix.1981 (J. Wolschrijn), gent. CG 2449).

277. *M. semiodactylus* (Mann) – France, Corsica, Ajaccio, St Antone, 80 m, 1.v.1988 (J. Nel), gent. CG 2340.

278. *M. hedemanni* (Rebel) – Canary Islands, Tenerife, 2 km S Vilaflor, 1100 m, 6.iv.1981 (F. Coenen), gent. CG 2348.

279. *M. chordodactylus* (Staudinger) – Canary Islands, Tenerife, Puerto de la Cruz, 25.iii-5.iv.1968 (B. van Aartsen), gent. CG 1426 (ITZ).

280. *M. bystropogonis* (Walsingham) Paralectotype – Canary Islands, Tenerife, Guimar, 28.iii-18.iv.1907 (Walsingham), gent. CG 5052 (BMNH).

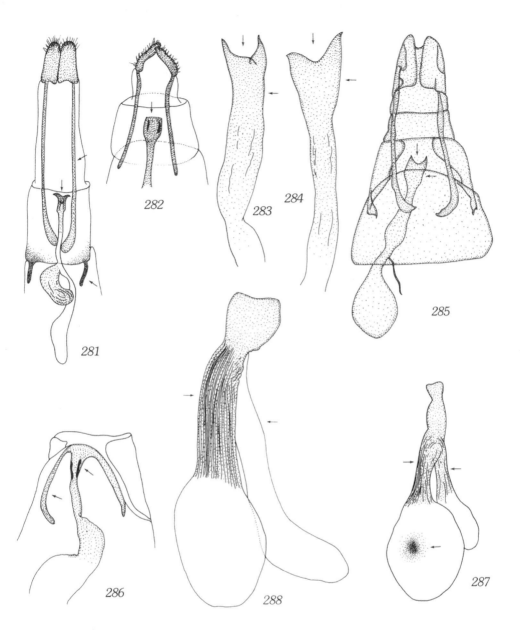

FEMALE GENITALIA

281. *Wheeleria phlomidis* (Staudinger) – Turkey, Konya, Sultan Daglari, Aksehir Pass, 1700 m, 4-23.vii.1974 (Groß), gent. CG 2344.
282. *W. raphiodactyla* (Rebel) – Spain, Granada, Sierra Nevada, Veleta Road, 2000 m, 23-29.vii.1986, gent. CG 1899.
283. *W. spilodactylus* (Curtis) – Great Britain, Wales, Gwynned, 5.vii.1984 (M. Hull), gent. CG 2352.
284. *W. obsoletus* (Zeller) – Austria, Nieder Österreich, Oberwiesen, 11.vii.1941 (Ortner), gent. CG 2353.
285. *W. lyrae* (Arenberger) – Greece, Lakonia, Waterfall near Nomia-Lyra, 17.v.1979 (E. Arenberger), gent. Arenberger 1965 (MNMB).
286. *W. ivae* (Kasy) – Turkey, Aksehir, 1000 m, 27-30.vi.1963 (Noach), gent. CG 2346.
287. *Pterophorus pentadactyla* (Linnaeus) – France, Isère, Freney d'Oisans, 1500 m, 6-28.vii.1987, gent. CG 2336.
288. *P. ischnodactyla* (Treitschke) – France, Vaucluse, Mont Ventoux, 9.iv.1989 (J. Nel), gent. CG 2347.

Distribution Catalogue

The presence of a given species in a European country is given by an "x", and also the presence in a number of non-European countries is indicated. Doubtful records which need confirmation are marked with a "?".

Acronyms for the countries

AB	Albania	IT	Italy
AL	Algeria	LS	Lebanon & Syria
AT	Austria	LU	Luxemburg
BA	Baltic Republics	LY	Libya
BE	Belgium	MD	Mediterrenean Isl.
BG	Bulgaria	MO	Morocco
CA	Canary Isl.& Madeira	NL	Netherlands
CZ	Czech Republic	NO	Norway
DK	Denmark	PO	Poland
EG	Egypt	PT	Portugal
FI	Finland	RO	Romania
FR	France	RU	Russia[1]
GB	Great Britain	SE	Sweden
GE	Germany	SK	Slovakia
GR	Greece	SP	Spain
HU	Hungary	SZ	Switzerland
IC	Iceland	TR	Turkey
IR	Ireland	TU	Tunesia
IS	Israel	YU	Yugoslavia[2]

[1] Russia: including Belorussia and Ukraine.
[2] Yugoslavia: comprises Slovenia, Croatia, Bosnia-Herzegovina, Macedonia, Montenegro and Serbia.

	A B	A L	A T	B A	B E	B G	C A	C Z	D K	E G	F I	F R	G B	G E	G R	H U	I C	I R	I S	L T	L S	L U	M Y	M D	N O	N L	P O	P O	R T	R O	S U	S E	S K	S P	T Z	T R	Y U
1. A. tamaricis	x			x			x		x		x		x		x		x	x					x								x	x	x	x	x		
2. A. intermedia									x																		x	x									
3. A. bennetii	x			x		x			x	x	x	x					x							x							x				x	x	
4. A. meridionalis	x			x			x ?										x			x	x										x					x	
5. A. salsolae			x																																		
6. A. delicatulella																	x																				
7. A. neglecta						x											x										x										
8. A. protai									x								x																				
9. A. adactyla		x			x		x		x	x	x		x	x		x											x		x	x	x						
10. A. heydeni			x	x			x			x			x	x		x	x	x				x					x			x	x						
11. A. satanas			x				x			x	x		x	x		x	x	x									x										
12. A. frankeniae	x			x		x		x	x		x						x	x		x	x	x			x		x	x		x	x	x					
13. A. gittia																											x										
14. A. espunae																							x				x										
15. A. glaseri																											x										
16. A. bigoti								x																													
17. A. symmetrica																	x										x										x
18. A. manicata								x									x	x					x		x		x										
19. A. paralia								x					x				x	x									x										x
20. A. bifurcatus			x																				x				x										
21. A. pseudocanariensis			x																				x				x										
22. A. hartigi																	x										x										
23. A. betica																											x										
24. P. tesseradactyla		x	x	x		x x		x x x x			x		x			x		x			x x x x															x
25. P. farfarellus	x		x	x				x		x x			x			x			x x x x																x
26. P. nemoralis		x	x x	x		x x			x			x			x			x	x													x
27. P. gonodactyla		x x x x	x x		x x x x x x x		x	x		x x		x		x x x x x x			x
28. P. calodactyla		x x x	x x		x x x x		x	x		x x x x		x x x x x		x
29. P. iberica																											x										
30. P. isodactylus					x		x x x		x			x x		x																					
31. G. miantodactylus			x				x		x				x			x x							
32. G. pallidactyla		x x x ?	x x		x x x x ? x		x	x	x		x x		x x						
33. G. tetradactyla	x	x x x x	x x		x x x x		x	x		x x		x x x		x
34. L. pusillidactylus			x	x													x																				
35. S. taprobanes	x		x x		x		x			x		x x x		x x x		x			x		x x
36. P. metzneri	x		x			x		x																			x				x
37. A. acanthodactyla	x x x x x x x x x		x x x x		x		x x x x x x x x x x		x x x x x x x x x x
38. A. punctidactyla		x x	x	x x		x x x x		x			x		x x x x x		x
39. S. graphodactyla	x		x	x		x	x		x					x ? x		x
40. S. pneumonanthes		x x	x		x	x				x			x x		x
41. S. gratiolae		x	x x		x	x		x					x x
42. S. pterodactyla	x	x x x x x x x	x x x x x x		x	x	x x x x		x x x x x x x		x
43. S. mannii		x								x			?			x x
44. S. veronicae: bipunctidactyla	x	x x x x ? x x ? x x x x x x		x x x x x		? x x x x		x x x x x x x x
45. S. sp group: bipunctidactyla																																					
46. S. sp group: aridus	? x ?		? ?		x	x		x ?		x		x x		x ? x x x
47. S. elkefi					x							x		x		x
48. S. lucasi					?													
49. S. annadactyla	x			x		x	x		x				x x x
50. S. sp group: pelidnodactyla	x x x x	x x		x x x	x		x	x	x		x
50a. S. sp group: brigantiensis					x												
51. S. reisseri																											x										
52. S. hahni																											x										
53. S. millieridactyla			x x		x	x		x
54. S. islandicus			x	x		x		x		x
55. S. parnasia						x												
56. S. coprodactylus	x	x	x		x	x x		x		x		x x x x		x
57. S. lutescens					x			x				x x
58. S. nepetellae					x									
59. S. stigmatodactylus	x x x	x x	x		x	x x x		x		x x x		x		x x x x x x x
60. S. stigmatoides																											x										
61. S. zophodactylus	x x x	x x	x x		x x x x x		x	x	x x x	x		x x x x x x x		x
62. B. capnodactylus		x	x x x		x	x		x			x		x x		x	x
63. C. rhododactyla	x	x x x x	x x		x x x x x x		x x x		x x		x x x x x x x
64. M. lunaedactyla		x x x		x x x x x		x	x		x	x		x x x x
65. M. fauna					x												
66. M. oxydactylus		x			x			x		x	x		x x
67. G. didactyla		x x	x	x x		x x	x		x		x		x x x x x		x
68. P. maculatus					x				x			x
69. P. croatica			?																																		x

200

	AB	AL	AT	BA	BE	BG	CA	CZ	DK	EG	FI	FR	GB	GE	GR	HU	IC	IR	IS	LT	LS	LU	MY	MD	NO	NL	PO	PO	RT	RO	SU	SE	SK	SP	TZ	TR	YU
70. P. anatolicus															x																			x		x	
71. C. britanniodactylus		x	x										x	x	x									x									x				
72. C. celeusi		x	x				x				x	x				x													x				x	x		x	x
73. C. trichodactyla	x	x	x	x	x						x	x																				x	x	x	x		x
74. C. fusca		x				x					x	x				x																x	x	x	x		x
75. C. bonneaui																																	x				
76. C. hellenica						x			x						x									x					x				x		x	x	x
77. C. loranus	x	x		x																													x				
78. C. marginellus																		x																			
79. C. zelleri																		x																			
80. C. polonica									x		x					x																					
81. C. maratonica		x							x				x	x		x							x				x					x				x	
82. B. paludum	x	x		x	x		x	x	x	x		x						x				x	x						x	x	x	x	x	x			
83. O. pilosellae	x	x	x	x	x	x	x	x	x	x		x			x		x		x		x	x	x			x	x	x	x	x	x	x	x	x			x
84. O. chrysodactyla	x	x	x	x	x	x	x	x	x	x					x		x				x	x			x	x	x	x	x	x	x	x				
85. O. ericetorum		x	x	x	x	x	x	x	x						x					x							x				x	x	x	x	x		x
86. O. parvidactyla	x	x	x	x	x	x	x	x	x	x	x	x			x				x	x	x				x	x	x		x	x	x	x	x	x	x	x	x
87. C. distans	x	x	x	x	x	x	x	x	x	x	x	x	x	x	x				x		x	x	x	x	x				x	x	x	x	x	x			x
88. C. tristis	x			x		x			x		x	x				x		x		x												x	x	x			
89. C. kollari	x								x							x																		x			
90. C. laetus		x	x				x	x			x					x			x	x	x				x					x			x			x	
91. S. siceliota		x	x				x				x					x	x	x		x	x				x					x	x	x	x	x	x		
92. M. leucodactylus											x																						x				
93. P. olbiadactylus			x				x				x					x	x	x		x	x				x					x			x				
94. G. baptodactylus							x									x				x													x				x
95. G. bigoti							x																														
96. G. nielswolffi			x																																		
97. P. heterodactyla	x	x	x	x		x	x		x		x	x	x	x	x			x		x		x			x	x			x	x			x	x		x	x
98. H. jnulae	x			x	x	x					x			x	x	x		x	x				x				x					x	x				x
99. H. sp group: carphodactyla	x	x	x	x		x					x	x	x					x	x		x		x						x	x	x						x
100. H. chrysocomae													x	x	x																						
101. H. osteodactylus	x	x		x			x	x			x	x	x	x				x		x	x	x		x	x				x	x			x	x	x	x	x
102. H. pectodactylus				x							x		x	x																					x		
103. H. distinctus		x				x	x		x	x		x				x								x	x							x	x	x	x		
104. H. didactylites	x	x	x	x		x	x		x	x		x				x		x					x	x						x	x	x	x	x	x	x	x
105. H. tephradactyla	x	x	x	x		x	x		x	x	x	x	x			x							x	x					x	x	x		x	x	x		
106. H. lienigianus	x	x	x	x		x	x		x	x	x	x	x	x	x			x											x	x			x	x			
107. O. lithodactyla	x	x	x	x		x	x		x	x	x	x			x			x		x					x				x	x	x	x	x	x	x		x
108. O. rogenhoferi	x								x							x							x						x	x			x	x			
109. O. constanti	x					x					x		x																				x	x			x
110. O. giganteus									x									x					x				x										
111. O. vafradactylus																																	x				
112. E. monodactyla	x	x	x	x	x	x	x		x	x	x	x	x	x	x		x	x	x	x	x	x	x	x	x			x	x	x	x	x	x	x	x	x	x
113. E. argoteles	x								x		x	x			x								x										x	x			
114. A. microdactyla	x	x	x	x		x	x		x	x	x	x	x	x				x		x		x	x	x					x	x	x	x	x	x	x		x
115. C. punctinervis									x							x									x								x				
116. C. homoiodactyla			x	x					x			x																					x		x		x
117. C. adamas									x							x									x								x				
118. C. acarnella																		x																			
119. C. albodactylus	x		x	x	x				x			x	x	x				x		x									x	x	x						x
120. C. xanthodactyla												x																					x				x
121. C. nephelodactyla	x			x					x			x				x													x		x	x	x		x		x
122. P. galactodactyla	x				x	x	x	x	x	x		x													x				x	x							
123. M. leucodactyla	x	x	x	x		x	x	x	x	x	x	x					x		x		x		x						x	x	x	x	x	x	x		x
124. M. tridactyla	x		x		x	x	x	x	x	x	x	x	x		x		x				x	x			x				x	x	x	x	x	x	x		x
125. M. baliodactylus	x	x	x	x		x	x		x	x			x	x				x	x						x					x			x	x	x	x	x
126. M. malacodactylus			x						x			x	x					x	x	x				x	x				x			x		x	x	x	x
127. M. semiodactylus																		x	x						x									x			
128. M. hedemanni																																					
129. M. chordodactylus				x																					x								x				
130. M. bystropogonis				x																																	
131. W. phlomidis								x				x	x												x								x				
132. W. raphiodactyla							x																										x				
133. W. spilodactylus		x							x	x	x	x					x		x	x													x			?	x
134. W. obsoletus	x		x		x				x			x	x				x	x	x		x	x					x	x		x			x			x	x
135. W. lyrae										x																											
136. W. ivae			x								x								x																	x	x
137. P. pentadactyla	x	x	x	x		x	x		x	x	x	x	x	x	x			x	x		x	x	x					x	x	x	x	x	x	x	x	x	x
138. P. ischnodactyla	x				x				x			x				x		x	x				x						x				x		x	x	x

References

Aarvik, L. *et al.*, 1986. New and interesting records of Lepidoptera from Norway.– *Fauna norv. Ser. B* **33**: 77-90.

Aarvik, L., 1987. Contribution to the knowledge of the Norwegian Lepidoptera, II.– *Fauna norv. Ser. B* **34**: 7-13.

Adamczewski, S., 1948. Notes on the plume-moths II. *Capperia trichodactyla* (Den. & Schiff., 1775) in Poland (Lep., Alucitidae).– *Polskie Pismo ent.* **18** (2-4): 142-155.

Adamczewski, S., 1951. On the systematics and origin of the generic group *Oxyptilus* Zeller (Lep., Alucitidae).– *Bull. Br. Mus. nat. Hist.* (Ent.) **1** (5): 301-388, plates 9-20.

Agenjo, R., 1952. Nuevo *Agdistis* Betico-Marroqui.– *Trans. IXth. Int. Congr. Ent. Amsterdam* **1**: 121-124.

Amsel, H.G., 1935a. Neue palästinensische Lepidopteren.– *Mitt. Zool. Mus. Berl.* **20**: 271-319.

Amsel, H.G., 1935b. Weitere Mitteilungen über palaestinensische Lepidopteren.– *Veröff. dt. Kolon. u. Übersee-Mus. Bremen* **1**: 225-277.

Amsel, H.G., 1949. Microlepidoptera collected by E. P. Wiltshire in Iraq and Iran.– *Bull. Soc. Fouad I. Ent.* **33**: 310-311.

Amsel, H.G., 1951a. Neue Maroccanische Kleinschmetterlinge.– *Bull. Soc. Sci. nat. Maroc* **31**: 65-74.

Amsel, H.G., 1951b. Lepidoptera Sardinica. III. Descrizione di specie nuove ed osservazioni-sistematiche di carattere generale.– *Fragm. ent.* **1**: 101-144.

Amsel, H.G., 1954a. Neue Pterophoriden, Gelechiiden und Tineiden aus Palästina und Malta (Microlep.).– *Bull. Soc. Fouad I. Ent.* **38**: 51-53.

Amsel, H.G., 1954b. Neue oder bemerkenswerte Kleinschmetterlinge aus Österreich, Italien, Sardinien und Corsica.– *Z. wien. ent. Ges.* **39**: 5-17.

Amsel, H.G., 1955a. Kleinschmetterlinge vom Jordantal.– *Z. wien. ent. Ges.* **40**: 276-282.

Amsel, H.G., 1955b. Über mediterrane Microlepidopteren und einige transcaspische Arten.– *Bull. Inst. r. Sci. nat. Belg.* **31** (83): 23-24, 51-57.

Amsel, H.G., 1956. Über die von Herrn E. de Bros in Spanisch-Marokko gesammelten Klein-Schmetterlinge.– *Z. wien. ent. Ges.* **41**: 17-31.

Amsel, H.G., 1959. Irakische Kleinschmetterlinge II.– *Bull. Soc. ent. Egypt* **43**: 54-56.

Amsel, H.G., 1968. Zur Kenntnis der Microlepidopteren Fauna von Karachi (Pakistan).– *Stuttg. Beitr. Naturk.* **191**: 14-15.

Arenberger, E., 1973a. Eine neue *Agdistis*-Art von den Kanarischen Inseln.– *Beitr. naturk. Forsch. Süds.Dtl.* **32**: 179-180.

Arenberger, E., 1973b. Eine neue *Agdistis*-Art aus Sardinien.– *Studi Sass. III* **21** (1): 1-16.

Arenberger, E., 1973c. Die *Agdistis*-Arten Sardiniens.– *Studi Sass. III* **21** (2): 641-654.

Arenberger, E., 1976a. Neue *Agdistis*-Arten.– *Dt. ent. Z.* (N.F.) **23**: 61-67.

Arenberger, E., 1976b. Eine *Agdistis*-Art von Kreta.– *Z. ArbGem. öst. Ent.* **28**: 7-8.

Arenberger, E., 1977. Die palaearktischen *Agdistis*-Arten.– *Beitr. naturk. Forsch. Süds.Dtl.* **36**: 185-226.

Arenberger, E., 1978. *Agdistis*-Arten aus Spanien.– *Z. ArbGem. öst. Ent.* **29**: 73-80.

Arenberger, E., 1981. Die *Pterophorus*-Arten West- und Zentralasiens. 1. Beitrag.– *Z. ArbGem. öst. Ent.* **32**: 97-110.

Arenberger, E., 1983. Records of the Lepidoptera of Greece based on the collections of G. Christensen and L. Gozmány: II Pterophoridae.– *Annls Mus. Goulandris* **6**: 199-206.

Arenberger, E., 1984. Neue palaeaktischen Pterophoridae.– Z. ArbGem. öst. Ent. **36**: 8-14.

Arenberger, E., 1986. Ergänzende Bemerkungen zur Familie Pterophoridae.– Z. ArbGem. öst. Ent. **37**: 76-80.

Arenberger, E., 1988a. Taxonomische Klarstellungen bei den Pterophoridae.– Stapfia **16**: 1-12.

Arenberger, E., 1988b. Weitere palaearktische Pterophoridae.– Z. ArbGem. öst. Ent. **39**: 65-70.

Arenberger, E., 1989. Stenoptilia hahni nov. sp. - ein Neufund aus Spanien.– SHILAP **17**: 327-331.

Arenberger, E., 1990a. Vorarbeiten für die "Microlepidoptera Palaearctica": Der Pselnophorus Komplex.– NachrBl. bayer. Ent. **39**: 13-20.

Arenberger, E., 1990b. Neufunde von Pterophoridae in Österreich.– Z. ArbGem. öst. Ent. **42**: 55-57.

Arenberger, E., 1990c. Beitrag zur Kenntniss der Gattung Stenoptilia Hübner, 1825.– Nota lepid. **13**: 90-107.

Arenberger, E. & Jaksic, P., 1991. Fauna Durmitora, Sveska 4. Pterophoridae.– Crnogorska Akad. nauk. umjetnosti Posebna izdanja **24 (15)**: 225-242.

Barnes, W. & Lindsey, A.W., 1921. The Pterophoridae of America, north of Mexico.– Contr. Nat. Hist. Lepidoptera North America **4**: 280-483, 14 plates.

Becker, L., 1861. Transformations du Pterophorus scarodactylus.– Annls Soc. ent. Belg. **5**: 55-56.

Beirne B.P., 1954. British Pyralid and Plume Moths: 1-208, 16 plates. London & New York.

Biesenbaum, W., 1987. Erstfund von Platyptilia capnodactyla Zeller, 1841 in Nordrhein-West-falen.– Mitt. ArbGem. rhein.-westf. Lepidopterologen **5**: 2-3.

Bigot, L., 1960. Un Agdistis nouveau de Tripolitaine: A. fiorii n. sp.– Alexanor **1**: 201-202.

Bigot, L., 1961. Une nouvelle variété d'Aciptilia: A. spicidactyla insularis nova.– Lambillionea **61**: 7-8.

Bigot, L., 1962. Les Pterophoridae des îles Seychelles.– Bull. Soc. ent. Fr. **67**: 79-88, 9 figs.

Bigot, L., 1963. Les Pterophoridae de l'île de Crête.– Lambillionea **63**: 9-37, 26 figs.

Bigot, L., 1972. Nouvelles données sur les Lépidoptères Pterophoridae des îles Canaries.– Bull. Soc. ent. Fr. **77**: 223-228.

Bigot, L., 1974. Une sous-espèce nouvelle d'Agdistis pavalia Zeller dans les iles du littoral de Marseille (Lep. Pterophoridae).– Bull. Soc. ent. Fr. **79**: 85-90.

Bigot, L., 1987. Une espèce de Capperia dans le nord-est de l'Espagne: C. bonneaui nova species.– Alexanor **15 Suppl.**: [35]-[37].

Bigot, L. et al., 1990. Nouvelles observations écologiques et chronologiques sur les Lépidoptères Pterophoridae du Mont Ventoux. Treizième contribution à l'étude du peuplement en Lépidoptères du Mont Ventoux.– Alexanor **16 Suppl.**: [3]-[46].

Bigot, L., Nel, J. & Picard, J., 1986. Découverte en Provence des imagos et des chenilles de Capperia maratonica Adamczewski, 1951, espèce nouvelle pour la faune française.– Alexanor **14 Suppl.**: [18]-[21].

Bigot, L., Nel, J. & Picard, J., 1989. Oxyptilus pravieli nova species.– Alexanor **16**: 15-21.

Bigot, L., Nel, J. & Picard, J., 1990. Oxyptilus gibeauxi nova species en forêt de Fontainebleau.– Bull. Ass. Nat. Vall. Loing Massif Fontainebleau **66**: 47-58.

Bigot, L., Nel, J., & Picard, J., 1993. Merrifieldia meristodactyla (Zeller, 1852) (Mann in litteris), stat. rev. Merrifieldia inopinata nov.sp. du sud-est de la France (Lepidoptera Pterophoridae).– Alexanor **18**: 117-128.

Bigot, L. & Picard, J., 1981. Redécouverte du biotope d'origine d'Oxyptilus lantoscanus Millière dans le Parc national du Mercantour et nouvelles stations de l'espèce dans le Parc national des Ecrins et aux limites du Parc naturel régional du Queyras.– Alexanor **12**: 67-69.

Bigot, L. & Picard, J., 1983a. Lépidoptères ptérophorides du département des Bouches-du-Rhône et de la région de la Sainte-Baume (Supplément au catalogue de P. Siépi).– *Bull. Mus. Hist. nat. Marseille* **43**: 53-69.

Bigot, L. & Picard, J., 1983b. Une espèce nouvelle de *Stenoptilia* dans le sud-est de la France: *S. nepetellae* n. sp.– *Alexanor* **13**: 21-24.

Bigot, L. & Picard, J., 1986a. *Paraplatyptilia* n. nov. pour *Mariana* Tutt, 1907, préoccupé. Nouvelle capture en France de *Stenoptilodes taprobanes* (Felder & Rogenhofer, 1875).– *Alexanor* **14 Suppl.**: [17].

Bigot, L. & Picard, J., 1986b. Notes sur les espèces européennes du genre *Capperia* et création de deux nouveax sous-genres.– *Alexanor* **14 Suppl.**: [21]-[24].

Bigot, L. & Picard, J., 1988a. Remarques sur les *Oxyptilus* (1e partie). Généralités, problèmes liés à *O. hieracii* (Zeller, 1841). Descriptions *d'O. buvati* et *d'O. adamczewskii*, nouvelles espèces.– *Alexanor* **15**: 239-248.

Bigot, L. & Picard, J., 1988b. Remarques sur les *Oxyptilus* (2e partie). Notes complémentaires sur la systématique et sur la répartition des espèces. Clé de détermination des espèces françaises.– *Alexanor* **15**: 249-256.

Bigot, L. & Picard, J., 1989. Remarques sur les Pterophoridae français du genre *Merrifieldia; M. neli* et *M. garrigae*, espèces nouvelles.– *Alexanor* **15 Suppl.**: [25]-[40].

Bigot, L. & Picard, J., 1991. Remarques sur les *Oxyptilus* (3e partie). Compléments et rectifications. Description *d'O. jaeckhi*, nouvelle espèce. Réflexions sur les espèces de la section *distans*. Modifications àpporter aux clés de determination.– *Alexanor* **17**: 233-246.

Bigot, L. & Popescu-Gorj, A., 1973. Les Pterophoridae de la collection du musée "Gr. Antipa" de Bucarest, I & II.– *Trav. Mus. Hist. nat. "Gr. Antipa"* **13**: 185-194; **14**: 221-232.

Bretherton, R.F., 1956. *Leioptilus carphodactylus* Hübner in North Surrey.– *Entomologist´s Rec. J. Var.* **68**: 272.

Bruand M. d'Uzelle, 1859. *In:* Laboulbène, A: Rapport sur la session extraordinaire tenue à Grenoble au mois de juillet 1858.– *Annls Soc. ent. Fr.* **27**: 819-904.

Bruand, M. d'Uzelle, 1861. Note sur quelques espèces du genre *Pterophorus*.– *Annls Soc. ent. Fr.* **30**: 33-38.

Büttner, F.O., 1880. Die Pommerschen, insbesondere die Stettiner Microlepidopteren.– *Stettin. ent. Ztg.* **41**: 471-473.

Buhl, O. *et al.*, 1983. Fund af småsommerfugle fra Danmark i 1983.– *Ent. Meddr.* **50**: 11-20.

Burmann, K., 1944. Ein kleiner Beitrag zur Lebenskunde und Verbreitung von *Pterophorus rogenhoferi* Mann.– *Z. wien. ent. Ges.* **29**: 276-283.

Burmann, K., 1950. Die Raupe und Puppe von *Oxyptilus kollari* Stainton.– *Z. wien. ent. Ges.* **35**: 146-147.

Burmann, K., 1954. *Stenoptilia pelidnodactyla* Stein n. ssp. *alpinalis*.– *Z. wien. ent. Ges.* **39**: 187-191.

Burmann, K., 1965. *Pterophorus nephelodactyla* Eversmann in den österreichischen Alpen.– *Z. wien. ent. Ges.* **50**: 66-68.

Burmann, K., 1986. Beiträge zur Microlepidopteren-Fauna Tirols. XI. Pterophoridae.– *Ber. nat.-med. Verein Innsbruck* **73**: 133-146.

Busck A., 1914. New microlepidoptera from Hawaii.– *Insecutor Insciti. menstr.* **2**: 103-104.

Buszko, J., 1974. *Aciptilia xanthodactyla* (Tr.) in Polen.– *Polskie Pismo ent.* **44**: 737-740.

Buszko, J., 1975. *Aciptilia exilidactyla* sp. n., a new species of plume moth from Central Europe.– *Polskie Pismo ent.* **45**: 141-146.

Buszko, J., 1978. Über systematische Stellung der Gattungen in der Gattungs-Gruppe *Stenoptilia-Platyptilia*.– *Polskie Pismo ent.* **48**: 67-79.

Buszko, J., 1979a. Pterophoridae Bulgariens.– *Polskie Pismo ent.* **49**: 683-703.

Buszko, J., 1979b. Der Verbreitungscharacter von Pterophoridae in Mitteleuropa.– *Symp. Int. Entomofaun. Europae Median* **7**: 248-250.

Buszko, J., 1986. A review of Polish Pterophoridae.– *Polskie Pismo ent.* **56**: 273-315.

Butler, A.G., 1882 (dated 1881). On a collection of nocturnal lepidoptera from the Hawaiian Islands.– *Ann. Mag. nat. Hist.* **7 (5th series)**: 407-408.

Cansdale, W.D., 1955. *Amblyptilia acanthodactyla* in Essex.– *Entomologist's Rec. J. Var.* **67**: 303.

Caradja, A., 1920. Beitrag zur Kenntnis der geographischen Verbreitung der Mikrolepidopteren des palaearktischen Faunengebietes, nebst Beschreibungen neuer Formen. III. Teil.– *Dt. ent. Z. Iris* **34**: 75-179.

Champion, H.G., 1910. Occurence of *Trichoptilus paludum* Zell. near Woking.– *Entomologist's mon. Mag.* **46**: 239-240.

Chapman, T.A., 1906a. *Marasmarcha agrorum* var. *tuttodactyla*, new var. (n. sp.).– *Entomologist's Rec. J. Var.* **18**: 178-179.

Chapman, T.A., 1906b. Observations on the life histories of *Trichoptilus paludum* Zeller.– *Trans. ent. Soc. London* **1906**: 133-154, 1 plate.

Chapman, T.A., 1908. On *Stenoptilia grandis* (sp. nov.).– *Trans. ent. Soc. London* **1908 (ii)**: 317-320; plates 14-17.

Chrétien, P., 1891. Description de micro-lépidoptères nouveaux.– *Naturaliste* **13 (2nd serie)**: 99.

Chrétien, P., 1917. Contribution à la connaissance des Lépidoptères du nord de l'Afrique, notes biologiques et critiques.– *Annls Soc. ent. Fr.* **85**: 462.

Chrétien, P., 1922. Les Lépidoptères du Maroc.– *Etud. lépidopt. comp.* **19**: 13-403, pls. 74-124.

Chrétien, P., 1923. Les chenilles des lavandes.– *Amat. Papillons* **1**: 229-235.

Chrétien, P., 1925. La légende de *Graellsia isabellae*.– *Amat. Papillons* **2**: 243.

Constant, M.A., 1865. Description de quelques Lépidoptères nouveaux.– *Annls Soc. ent. Fr.* **45**: 193.

Constant, M.A., 1885. Notes sur quelques lépidoptères nouveaux, 3ième partie.– *Annls Soc. ent. Fr.* **50**: 5-46

Constant, M.A., 1895. Microlépidoptères nouveaux de la faune française.– *Bull. Soc. ent. Fr.* l-liv.

Curtis, J., 1823-1840. *British Entomology: Being illustrations and descriptions of the genera of insects... etc.–* Volume 1-16. London.

Dale, J.C., 1864. Notes on the Pterophori with a description of *P. similidactylus* Curt.– *Mag. nat. Hist.* **7**: 263-264.

[Denis, J.N.C.M. & I. Schiffermüller], 1775. *Ankündigung eines Werkes von den Schmetterlingen der Wienergegend.* Pp. 1-322, 2 plates. Bernardi, Wien.

Derra, G., 1987. *Emmelina jezonica pseudojezonica* ssp. nov.– *Nota lepid.* **10**: 71-78.

Doets, C., 1946. Lepidopterologische mededelingen over 1939-45.– *Ent. Ber., Amst.* **12**: 84-91.

Doets, C., 1950. Notes on Lepidoptera, 1949.– *Ent. Ber., Amst.* **13**: 163-167.

Doets, C., 1952. Lepidopterologische mededelingen over 1950-1951.– *Ent. Ber., Amst.* **14**: 177-181.

Duponchel, P.A.J., 1840a. *In:* Godart, J.B.: *Histoire naturelle des Lépidoptères ou Papillons de France* **8**: 631-685. Méquigong-Marvis, Paris.

Duponchel, P.A.J., 1840b. *In:* Godart, J.B.: *Histoire naturelle des Lépidoptères ou Papillons de France* **11**: 720 pp., pls. 287-314. Méquignon-Marvis, Paris.

Duponchel, P.A.J., 1844. *In:* Godart, J.B.: *Histoire naturelle des Lépidoptères ou Papillons de*

France. Nocturnes, **Suppl. 4**: 534 pp., pls. 51-90. Méquignon-Marvis, Paris.

Dyar, H.G., 1903. List of Lepidoptera taken at Williams, Arizona, by Messrs. Schwarz and Barber.– 1. Papilionoidea, Sphingoidea, Bombycoidea, Tineioidea (in part).– *Proc. ent. Soc. Wash.* **5**: 228.

Emmet, A.M. (ed.), 1979. *A field guide to the smaller British Lepidoptera.* 1-271. London.

Emmet, A.M., 1983. The early stages of Crombrugghia distans (Zeller).– *Entomologist's Rec. J. Var.* **95**: 15-18.

Eversmann, E., 1844. *Fauna Lepidopterologica Volgo-Uralensis.* Casani.

Fabricius, J.C., 1794. *Entomologia systematica emendata et aucta.* Vol. **III**, pars **II**: 1-349. Hafniae, C. G. Proft.

Fazekas, I., 1986. Pterophorus malacodactylus transdanubinus n. ssp., eine neue Federmotten-Unterart aus Hungarn.– *Ent. Z., Frankf. a. M.* **96**: 12-16.

Felder, R. & Rogenhofer, A.F., 1875. Atlas der Heterocera. *In: Reise der Österreichischen Fregatte Novara um die Erde in den Jahren* 1857-1859. *Zoology* **2 (2)**: plate 140.

Fernald, C.H., 1898. The Pterophoridae of North America.– *Hatch Exped. Stn Mass. Agric. Coll.*: 1-83. Boston.

Fish, C., 1881. Pterophoridae.– *Can. Ent.* **13**: 70-74, 140-143.

Fitch, A., 1854. The gartered or grape-vine plume.– *Trans. N. Y. St. agric. Soc.* **14**: 843-849.

Fletcher, T.B., 1907. Description of a new plume-moth from Ceylon, with some remarks upon its life history.– *Spolia zeylan.* **5**: 20-32.

Fletcher, T.B., 1909. The plume-moths of Sri Lanka.– *Spolia Zeylanica* **6**: 1-39, 5 pl.

Fletcher, T.B., 1921 (1920). Life-Histories of Indian Insects, Microlepidoptera.– *Mem. Dep. Agri. India* (Entomological Series) **6**: i-ii, 1-217, plates 1-68. (Dated Nov. 1920, but correct date is Jan. 1921).

Fletcher, T.B. & Pierce, F.N., 1940. A new Irish plume-moth, with a note on its genitalia.– *Entomologist's Rec. J. Var.* **52**: 25-29.

Frey, H., 1856. *Die Tineen und Pterophoren der Schweiz.* i-xi, 1-430. Zürich.

Frey, H., 1870. Ein Beitrag zur Kenntniss der Microlepidopteren.– *Mitt. schweiz. ent. Ges.* **3**: 277-296.

Frey, H., 1886. Einige Micros aus Regensburg.– *Stettin. ent. Ztg.* **47**: 16-18.

Fuchs, A., 1895. Kleinschmetterlinge der Loreley-Gegend. 4te Besprechung.– *Stettin. ent. Ztg.* **56**: 21-52.

Fuchs, A., 1901. Bemerkungen zu zwei Nassauischen Pterophoriden.– *Jb. nassau. Ver. Naturk.* **54**: 70-72.

Fuchs, A., 1902. Neue Geometriden und Kleinfalter des europäischen Faunengebiets.– *Stettin. ent. Ztg.* **63**: 317-330.

Gaj, A.J., 1959. Notes on Pterophoridae. Platyptilia metzneri Zeller and related species.– *Ent. Ber., Amst.* **19**: 150-158.

Gartner, A., 1862. Lepidopterologische Beiträge.– *Wien. Ent. Monatschr.* **6**: 328-332.

Gibeaux, C., 1985. Révision des Stenoptilia de France avec la description de deux nouvelles (1e note).– *Ent. gall.* **1**: 237-265.

Gibeaux, C., 1986. Révision de quelques types; S. elkefi Arenberger en France; description de taxa nouveaux dans le groupe bipunctidactyla. Etude des Stenoptilia français (3e note).– *Alexanor* **14**: 323-335.

Gibeaux, C., 1989a. Etude des Pterophoridae (11e note). Une très belle découverte à Fontainebleau: Stenoptilia annickana n. sp.– *Bull. Ass. Nat. Vall. Loing Massif Fontainebleau* **64**: 222-229.

Gibeaux, C., 1989b. *Agdistis manicata* Stgr., espèce nouvelle pour la faune de France.– *Alexanor* **15 Suppl.**: [11]-[13].

Gibeaux, C., 1989c. Etude des Pterophoridae (8e note). Description d'un *Stenoptilia* nouveau dans le groupe *graphodactyla* Treitschke.– *Alexanor* **15 Suppl.**: [13]-[19].

Gibeaux, C., 1989d. Etude des Pterophoridae (12e note). Description de *Leioptilus inulaevorus* n. sp.– *Alexanor* **16**: 73-76.

Gibeaux, C., 1990a. Etude des Pterophoridae (16e note). Les types de Pterophoridae du Muséum national d'Histoire naturelle de Paris décrits jusqu'en 1964.– *Ent. gall.* **2**: 51-68.

Gibeaux, C., 1990b. Etude des Pterophoridae (23e note). Description de *Stenoptilia arenbergeri* n. sp., taxon du groupe *graphodactyla* (Treitschke, 1833).– *Bull. Ass. Nat. Vall. Loing Massif Fontainebleau* **66**: 219-225.

Gibeaux, C., 1990c. Etude des Pterophoridae (19e note). Description d'un ptérophore normand: *Capperia sequanensis* n. sp.– *Ent. gall.* **2**: 73-74.

Gibeaux, C., 1992. Etude des Pterophoridae (27e note). Caractérisation de taxa nouveaux dans le genre *Stenoptilia* entraînant la création d'une section *coprodactyla* Stainton.– *Bull. Soc. ent. Fr.* **96**: 463-471.

Gibeaux, C. & Nel, J., 1990a. Description d'une espéce nouvelle du genre *Gypsochares* Meyrick, 1890.– *Alexanor* **16**: 121-128.

Gibeaux, C. & Nel, J., 1990b. Description de *Stenoptilia gratiolae* n. sp. Etude des Pterophoridae (14e note).– *Bull. Ass. Nat. Vall. Loing Massif Fontainebleau* **65**: 199-209.

Gibeaux, C. & Nel, J., 1991. Révision des espèces françaises du complexe *bipunctidactyla* (Scopoli, 1763) dans le genre *Stenoptilia* Hübner, 1825.– *Alexanor* **17**: 103-119.

Gibeaux, C. & Picard, J., 1992. Les espèces françaises du genre *Oidaematophorus* Wallengren, 1862 (*Leioptilus auct.* inclus). Généralités. Inventaire systématique. *Oidaematophorus alpinus* nov. sp.– *Ent. gall.* **3**: 113-124.

Gielis, C., 1986. Quelques notes sur l'élevage et la repartition de *Stenoptilia nepetellae* Bigot et Picard, 1983.– *Alexanor* **14 Suppl.**: [3]-[6].

Gielis, C., 1993. Generic revision of the superfamily Pterophoroidea (Lepidoptera).– *Zool. Verh. Leiden* **290**: 1-139.

Gielis, C. & Arenberger, E., 1992. *Gypsochares nielswolffi* n. sp. from Madeira.– *Ent. Ber., Amst.* **52**: 81-84.

Goury, G., 1912. Observations sur la chenille de *Stenoptilia zophodactyla* Dup. Moers.- Hibernation.- Premiers états.– *Feuille jeune natural:* 174-177.

Gozmány, L.A., 1959. The results of the zoological collecting trip to Egypt in 1957, of the Natural History Museum, Budapest. 6. Egyptian Microlepidoptera. Part I.– *Ann. Hist. Nat. Mus. Nat. Hungar.* **51**: 367-368.

Gozmány, L.A., 1962. Microlepidoptera IV.– *Fauna Hungariae* **7**: 1-298, 135 figs. Budapest.

Graaf, H.W. de, 1859. Eene diagnostische beschrijving gemaakt van europesche Pterophoridae.– *Tijdschr. voor Entomologie* **2**: 35-57.

Graaf, H.W. de, 1868. Microlepidoptera.– *Tijdschr. voor Entomologie* **11**: 71-81.

Gregson, C.S., 1867. *Oxyptilus britanniodactylus*, a new plume.– *Entomologist* **4**: 305-306.

Gregson, C.S., 1868. A hitherto unpublished description of a new *Pterophorus*.– *Entomologist´s mon. Mag.* **4**: 178-179

Gregson, C.S., 1869. Captures of Lepidoptera in Westmoreland.– *Entomologist´s mon. Mag.* **6**: 115.

Gregson, C.S., 1871. *Pterophorus scabiodactylus*, Gregson, a new British plume.– *Entomologist* **4**: 363-364.

Gregson, C.S., 1885. *Mimaesoptelus scabiodactylus.*– *Entomologist* **18**: 151.

Grinnell, F., 1908. Notes on the Pterophoridae or plume-moths of Southern California, with descriptions of new species.– *Can. Ent.* **19**: 313-321.

Haggett, G., 1956. Unusual foodplant of *Stenoptilia bipunctidactyla* Scopoli.– *Entomologist's Gaz.* **7**: 40.

Hannemann, H.J., 1976. Notizen über Federmotten.– *Dt. ent. Z.* (N.F.) **23**: 295-296.

Hannemann, H.J., 1977a. Über *Platyptilia capnodactyla* (Zeller, 1841).– *Dt. ent. Z.* (N.F.) **24**: 219-221.

Hannemann, H.J., 1977b. Kleinschmetterlinge oder Microlepidoptera, III. Federmotten (Pterophoridae), Gespinstmotten (Yponomeutidae), Echte Motten (Tineidae).– *Tierwelt Dtl.* **63**: 1-274, 17 plates. Jena.

Hartig, F., 1953. Descrizione di tre specie di lepidotteri dell'isola de Zannore.– *Boll. Soc. ent. Ital.* **83**: 67-69.

Haworth, A.H., 1811. *Lepidoptera Britannica* III: 377-511 (1812). J. Murray, London.

Heckford, R.J., 1988. Discovery in England of the larva of *Pterophorus fuscolimbatus phillipsi* Huggins.– *Entomologist's Gaz.* **39**: 189-191.

Hering, E., 1891. Ergänzungen und Berichtigungen zu F.O. Büttner's Pommerschen Mikrolepidopteren.– *Stettin. ent. Ztg.* **52**: 135-227.

Herrich-Schäffer, G.A.W., 1847-1855. *Systematische Bearbeitung der Schmetterlinge von Europa, zugleich als Text, Revision und Supplement zu Jakob Hübner's Sammlung europäischer Schmetterlinge.* Band 5, Die Schaben und Federmotten. 1-394. Regensburg.

Hochenwarth, S., 1785. Beiträge zur Insectengeschichte.– *Sch. berl. Ges. Naturf. Berl.* **6**: 334-360.

Hofmann, O., 1896. Die deutschen Pterophorinen, systematisch und biologisch bearbeitet.– *Ber. naturw. Ver. Regensburg* **5**: 25-219, 3 plates.

Hofmann, O., 1898a. Beobachtungen über die Naturgeschichte einiger Pterophoriden-Arten.– *Illte. Z. Ent.* **3**: 306-308, 339-342.

Hofmann, O, 1898b. Eine neue *Amblyptilia*.– *Dt. ent. Z. Iris* **1898**: 33-34.

Hübner, J., 1796-[1834]. *Sammlung europaeischer Schmetterlinge.* 78 pp., 71 pls. Augsburg.

Hübner, J., 1816-[1826]. *Verzeichniss bekannter Schmettlinge (sic).* 431 pp. Augsburg.

Huggins, H.C., 1939. A few notes on *Platyptilia tesseradactylus.– Entomologist* **72**: 177-178.

Huggins, H.C., 1955. The Irish subspecies of *Alucita icterodactyla* Mann.– *Entomologist's Gaz.* **6**: 124-126.

Jäckh, E., 1936. Bemerkungen über *Trichoptilus paludum* Zeller.– *Mitt. ent. Ges. Halle* **14**: 5-7.

Jäckh, E., 1961. *Pterophorus nephelodactyla* Eversmann auch in den Italienischen Alpen.– *Boll. Soc. ent. Ital.* **91**: 9-10.

Janmoulle, E., 1939. A propos de la découvert d'*Agdistis bennetii* Curtis au Zwyn.– *Bull. Mus. r. Hist. nat. Belg.* **15 (62)**: 1-4.

Jordan, R.C.R., 1869. *Pterophorus hieracii.– Entomologist's mon. Mag.* **6**: 14-15.

Karsholt, O. & C. Gielis, 1995. The Pterophoridae described by J.C. Fabricius, with remarks on type material of Fabrician Lepidoptera.– *Steenstrupia* **21**: 31-35.

Karvonen, V.J., 1932. Vier neue Kleinschmetterlinge aus Finland.– *Notul. ent.* **12**: 79-81.

Kasy, F., 1960. *Calyciphora*, ein neues Subgenus; *klimeschi, ivae, homoiodactyla*, drei neue Arten des Genus *Aciptilia* Hb.– *Z. wien. ent. Ges.* **45**: 174-187, 1 plate.

Koçak, A.Ö., 1981. *Buszkoiana* nom. nov. A Replacement Name in the Family Pterophoridae.– *Priamus* **1**: 10.

Kollar, V., 1832. Systematisches Verzeichnis der Schmetterlinge im Erzherzogthum Österreich.– *Beitr. Landesk. Österr.* **2**: 1-101.

Kyrki, J. & Karvonen, J., 1985. *Calyciphora xerodactyla* in Finland.– *Notul. ent.* **65**: 106-108.

Latreille, P.A., 1796. *Précis des caractères génériques des insectes, disposés dans un ordre naturel.* i-xiii, 1-201. Paris.

Leech, J.H., 1886. *British pyralides including the Pterophoridae.* Pp. i-viii, 1-122, 18 plates. London.

Legrand, H., 1936. *Le Pterophorus giganteus* Mann.– *Amat. Papillons:* 78-82, 1 plate.

LHomme, L., 1939. XXII Pterophoridae. Pp. 174-202, in *Catalogue des Lépidoptères de France et de Belgique.* **2** (2): 173-307.

Linnaeus, C., 1758. *Systema Naturae per regna tria naturae, secundum classes, ordines genera, species, cum characteribus, differentiis, synonymis, locis.* Ed. X, (1), 534 pp. Laurentii Salvii, Holmiae.

Linnaeus, C., 1761. *Fauna Svecica.* [43] + 578 pp., 2 pls. Stockholmiae.

Lucas, D., 1955. Nouveaux lépidoptères Nord-Africains.– *Bull. Soc. Sci. nat. phys. Maroc* **15**: 251-258.

Mann, J., 1855. Die Lepidoptera gesammelt auf einer entomologischen Reise in Corsika im Jahre 1855.– *Verh. zool.-bot. Ges. Wien* **5 Abh.**: 529-572.

Mann, J., 1871. Beitrag zur Kenntnis der Lepidopteren-Fauna des Glockner-Gebietes, nebst Beschreibung drei neuer Arten.– *Verh. zool.-bot. Ges. Wien* **21**: 71-82.

Marek, J. & Skyva, J., 1985. Faunistic records from Czechoslovakia. Lepidoptera: Pterophoridae.– *Acta ent. bohemoslovaca* **82**: 394-395.

Matsumura, S., 1931. *6000 illustrated insects of Japan-Empire.* ii + ii + iii + 23 + 1497 + 191 + 2 + 6 pp., 10 pls. Tokohshoin, Tokyo.

Matthews, D.L., 1989. *The plume moths of Florida.* i-xv, 1-347 pp. University of Florida, Gainesville, Florida.

Mellini, E., 1954. *Pterophorus microdactylus* Hbn. nella biocenosi di *Eupatorium cannabinum.*– *Boll. Ist. Ent. Univ. Bologna* **20**: 275-307.

Meyrick, E., 1886. On the classification of the Pterophoridae.– *Trans. ent. Soc. London* **1886**: 1-21.

Meyrick, E., 1888. On the Pyralidina of the Havaiian Islands.– *Trans. ent. Soc. London* **1888**: 209-246.

Meyrick, E., 1890. On the classification of the Pyralidina of the European fauna.– *Trans. ent. Soc. London* **1890**: 429-492.

Meyrick, E., 1891. A fortnight in Algeria with descriptions of new lepidoptera.– *Entomologist's mon. Mag.* **27**: 9-13.

Meyrick, E., 1902. A new European species of Pterophoridae.– *Entomologist's mon. Mag.* **38**: 217.

Meyrick, E., 1907. A new European species of Pterophoridae.– *Entomologist's mon. Mag.* **43**: 146-147.

Meyrick, E., 1908. Lepidoptera Heterocera (Pyrales): Fam. Pterophoridae. *In:* Wytsman, P. (ed.): *Genera Insect.* **100**: 1-22, 1 plate. Tervueren.

Meyrick, E., 1913a. *Exot. Microlepidopt.* **1** (4): 97-128.

Meyrick, E., 1913b. Pterophoridae, Orneodidae. *In:* Wagner, H. (ed.): *Lepid. Cat.* **17**: 1-44.

Meyrick, E., 1921. *Exot. Microlepidopt.* **2** (14): 417-448.

Meyrick, E., 1922. *Exot. Microlepidopt.* **2** (18): 545-576.

Meyrick, E., 1923. Three new microlepidoptera from Cyprus.– *Entomologist* **56**: 277-278.

Meyrick, E., 1924. Exot. Microlepidopt. **3** (3): 65-96.

Meyrick, E., 1926. Microlepidoptera from the Galapagos Islands and Rapa.– *Trans. ent. Soc. London* **74**: 276.

Meyrick, E., 1927. *A revised handbook of British lepidoptera.* 1-914. Reprint, 1970.

Meyrick, E., 1930. *Exot. Microlepidopt.* **3 (18)**: 545-576.

Meyrick, E., 1937. Microlepidoptera exkl. Pyralidae. *In:* Caradja & Meyrick: Materialien zu einer Mikrolepidopterenfauna des Yülingshanmassivs.– *Dt. ent. Z. Iris* **51**: 169-182.

Michaelis, H.N., 1986. Species of Pyralidae and Pterophoridae (Lep.) in North Wales.– *Entomologist's Rec. J. Var.* **98**: 231-240.

Millière, P., 1854. Description de nouvelles espèces de Microlépidoptères.– *Annls Soc. ent. Fr.* (3) **2**: 59-68, pl. 3.

Millière, P., 1859-1874. *Iconographie et Description de Chenilles et Lépidoptères inédits.–* **1**, 424 pp, 50 pls (1859-64); **2**, 506 pp, 50 pls (1864-68); **3**, 488 pp, 54 pls (1869-74). Paris.

Millière, P., 1871-76. *Catalogue raisonné des Lépidoptères du Département des Alpes-Maritimes.–* **1**: 1-135 (1871); **2**: 137-247 (1873); **3**: 249-455, pls. I-II (1876). Cannes.

Millière, P., 1875-76. Description de Chenilles et de Lépidoptères d'Europe (2ᵉ partie).– *Bull. Soc. ent. Fr.* **59**: 166-168.

Millière, P., 1883. Lépidoptérologie.– *Annls Soc. linn. Lyon* (N.S.) **29**: 153-179, 181-188, pls. I-IV.

Mitterberger, K., 1912. Die Nahrungspflanzen der deutschen Federmotten-Raupen.– *Arch. Naturgesch.* **A 11**: 116-125.

Möschler, H.B., 1866. Aufzählung der in Andalusien 1865 von Herrn Graf v. Hoffmannsegg gesammelten Schmetterlinge.– *Berl. ent. Z.* **10**: 136-146.

Moore, F., 1887. *In:* Walsingham, M.A.: *The Lepidoptera of Sri Lanka*, **3**: 526-529.

Morris, M.G., 1963. *Pterophorus galactodactylus* (Schiff.) and *Platyptilia gonodactyla* (Schiff.) in Huntingdonshire.– *Entomologist's Gaz.* **14**: 1.

Mühlig, G.G., 1863. Eine neue Pterophoride, *Platyptilia dichrodactylus.–Stettin. ent. Ztg.* **24**: 213-214.

Müller, O.F., 1764. *Fauna Insectorum Fridrichsdalina.* 24 + 96 pp. Hafniae et Lipsiae, Gleditsch.

Müller-Rutz, J., 1914. *In:* Müller-Rutz, J. & Vorbrodt, K.: *Die Schmetterlinge der Schweiz* **2**: 1-727. Berlin.

Müller-Rutz, J., 1934. Über Microlepidopteren.– *Mitt. schweiz. ent. Ges.* **16**: 118-119.

Murray, D.P., 1957. The life history of *Alucita galactodactyla* Hbn.– *Entomologist's Rec. J. Var.* **69**: 231-232, 1 plate.

Nel, J., 1984. Serre-Eyrauds (Hte-Alpes) du 1er au 8 juin 1984 (Lep. Pter., Nymphal., Lycaen.).– *Ent. gall.* **1**: 168.

Nel, J., 1986a. Sur les premiers états des *Oidaematophorus* français. Première contribution á la connaissance de la biologie des Pterophoridae du sud de la France.– *Alexanor* **14 Suppl.**: [7]-[16].

Nel, J., 1986b. Sur les premiers états des *Procapperia* et des *Capperia* en Provence. Deuxième contribution á la connaissance de la biologie des Pterophoridae du sud de la France.– *Alexanor* **14 Suppl.**: [25]-[32].

Nel, J., 1986c. Sur les premiers états des *Cnaemidophorus, Marasmarcha, Geina* et *Stangeia* français. Troisième contribution á la connaissance de la biologie des Pterophoridae du sud de la France.– *Alexanor* **14 Suppl.**: [33]-[40].

Nel, J., 1986d. Note sur les *Stenoptilia* français des Saxifrages. Quatrième contribution á la connaisance des premiers états des Pterophoridae.– *Alexanor* **14 Suppl.**: [41]-[45].

Nel, J., 1987a. Sur les premiers états des *Pterophorus* de France. Cinquième contribution á la connaisance de la biologie des Pterophoridae du sud de la France.– *Alexanor* **15 Suppl.**: [29]-[34].

Nel, J., 1987b. Sur les premier états de divers *Stenoptilia* souvent confondus sous le nom de *S. bipunctidactyla* (Scopoli, 1763). Sixième contribution á la connaissance de la biologie des Pterophoridae de sud de la France.– *Alexanor* **15 Suppl.**: [45]-[58].

Nel, J., 1987c. Sur les premiers états des *Gypsochares* Meyrick, 1890, et des *Pselnophorus* Wallengren, 1881. Septième contribution á la connaisance de la biologie des Pterophoridae du sud de la France.– *Alexanor* **15 Suppl.**: [59]-[64].

Nel, J., 1988a. Sur les premiers états de *Oxyptilus* Zeller, 1841, français. Huitième contribution á la connaisance de la biologie des Pterophoridae du sud de la France.– *Alexanor* **15**: 283-302.

Nel, J., 1988b. Sur les premiers états des *Pterophorus* du sous-genre *Calyciphora* Kasy, 1960, de la France. Neuvième contribution á la connaisance de la biologie des Pterophoridae du sud de la France.– *Alexanor* **15**: 303-310.

Nel, J., 1988c. *Capperia britanniodactyla* (Gregson, 1869) en Provence et dans le Massif Central. Dixième contribution á la connaisance de la biologie des Pterophoridae du sud de la France.-- *Alexanor* **15**: 313-317.

Nel, J., 1989a. Rectificatif à propos des premiers états des divers *Stenoptilia* souvent confondus sous le nom de *S. bipunctidactyla* (Scopoli, 1763).– *Alexanor* **15 Suppl.**: [1]-[2].

Nel, J., 1989b. Quelques données biologiques sur les *Amblyptilia* Hübner, 1825, de France. Douzième contribution à la connaissance de la biologie des Pterophoridae de la France.– *Alexanor* **15 Suppl.**: [20]-[24].

Nel, J., 1989c. Notes sur les Ptérophores de la Corse. Description de *Stenoptilia gibeauxi* n. sp. 13e contribution à la connaisance de la biologie des Pterophoridae du sud de la France.– *Alexanor* **15**: 467-478.

Nel, J., 1989d. Sur les premiers états des *Merrifieldia* Tutt, 1905, de France. 14e contribution à la connaisance de la biologie des Pterophoridae du sud de la France.– *Alexanor* **15**: 487-503.

Nel, J., 1991. Deuxième note sur les Pterophoridae de la Corse. *Stenoptilia cyrnea* n. sp. et *Merrifieldia moulignieri* n. sp. 21ième contribution à la connaisance de la biologie des Pterophoridae du sud de la France.– *Alexanor* **17**: 167-182.

Nel, J. & Gibeaux, C., 1990. Les *Stenoptilia* inféodés aux saxifrages. I. Révision des taxa décrits et caractérisation d'espèces nouvelles dans le groupe *pelidnodactyla* (Stein, 1837).– *Ent. gall.* **2**: 131-150.

Nel, J. & Gibeaux, C., 1992. Les *Stenoptilia* inféodés aux saxifrages. II *S. brigantiensis* et *S. buvati*, espèces nouvelles.– *Ent. gall.* **3**: 53-57.

Nel, J. & Prola, C., 1989. Sur deux Ptérophores méditerranéens méconnus: *Stenoptilodes taprobanes* (Felder & Rogenhofer, 1875) et *Capperia hellenica* Adamczewski, 1951. Onzième contribution à la connaissance de la biologie des Pterophoridae du sud de la France.– *Alexanor* **15 Suppl.**: [3]-[10].

Nel, J. & Prola, C., 1991. *Emmelina pseudojezonica* Derra, 1987, stat. rev. Description de ses premiers états.– *Alexanor* **17**: 23-29.

Nielsen, P.K., 1962. Nogle iagttagelser over fjermøllet *Platyptilia capnodactyla* Zell.– *Flora Fauna Silkeborg* **68**: 74-76.

Opinion 703. *Pterophorus* Schäffer, 1766 (Insecta, Lepidoptera): Addition to the official list of generic names.– *Bull. zool. Nom.* **21**: 113-115.

Osthelder, L., 1939. Die Schmetterlinge Südbayerns und der angrenzenden nördlichen Kalkalpen.– *Mitt. münch. ent. Ges.* **29 Suppl.**: [1-112, 2 plates].

Packard, A.S., 1873. Catalogue of the Pyralidae of California, with descriptions of new Californian Pterophoridae.– *Ann. Lyceum nat. Hist.* **10**: 257-267.

Pagenstecher, A., 1900. Die Lepidopterenfauna des Bismarck- Archipels.– *Zoologica, Stuttg.* **27**: 238-241.

Parrella, M.P. & Kok, L.T., 1978. Bionomics of *Oidaematophorus monodactyla* on hedge bindweed, in Southwestern Virgina.– *Ann. ent. Soc. Am.* **71**: 1-4.

Peyerimhoff, H. de, 1875. Chasses lépidoptèrologiques en Auvergne et diagnose d'une espèce nouvelle.– *Petites Nouv. Ent.* **7**: 515-516.

Pierce, F.N. & Metcalfe, J.W., 1938. *The genitalia of the British Pyrales with the Deltoids and Plumes.* i-xiii, 1-67, 29 plates. Liverpool.

Porritt, G.T., 1875. Description of the larva of *Pterophorus rhododactyla.*– *Entomologist's mon. Mag.* **12**: 88-89.

Purdey, W., 1910. Note on the early stages of *Oxyptilus pilosellae.*– *Entomologist* **43**: 89-90.

Ragonot, E.L., 1875.– *Bull. Soc. ent. Fr.* **5**: 74, 113, 205, 230-231.

Rebel, H., 1896. Dritter Beitrag zur Lepidopteren-Fauna der Canaren.– *Annln naturh. Mus. Wien* **11**: 114-116.

Rebel, H., 1900. Neue palaearktische Tineen.– *Dt. ent. Z. Iris* **13**: 161-188.

Rebel, H., 1904. Studien über die Lepidopteren-Fauna der Balkanländer. II.– *Annln naturh. Mus. Wien* **19**: 323-326.

Rebel, H., 1912. Einige für die Lepidopterenfauna Österreich-Ungarns neue Arten.– *Verh. zool.-bot. Ges. Wien* **62**: 107.

Rebel, H., 1935. Neue Pterophoriden und Tineen aus der Sierra de Gredos (Kastilien).– *Z. öst. EntVer.* **20**: 9-11.

Retzius, A.J., 1783. *Caroli de Geerii genera et species Insectorum.*

Ridder, L. de, 1986. From a synonym to a subspecies: *Agdistis heydeni canariensis* Rebel, 1896.– *Phegea* **14**: 81-87.

Roessler, A., 1864. Über die neue neben *Platyptilus ochrodactylus* H.-S. einzureihende Art.– *Wien. ent. Monatschr.* **8**: 53-54.

Roessler, A., 1881. Die Schuppenflügler (Lepidopteren) des Kgl. Regierungsbezirkes Wiesbaden und ihre Entwicklungsgeschichte.– *Jb. nassau. Ver. Naturk.* **33+34**: 1-392.

Rothschild, N.C., 1913. Some notes on *Platyptilia miantodactyla.*– *Entomologist's mon. Mag.* **49**: 159-160.

Schäffer, J.C., 1766. *Elementa Entomologica.* Pls. 1-135. Ratisbonae.

Schawerda, K., 1913. Siebenter Nachtrag zur Lepidopterenfauna Bosniens und der Herzegowina.– *Verh. zool.-bot. Ges. Wien.* **63**: 141-178.

Schawerda, K., 1933. Meine achtste Lepidopterenausbeute aus dem Hochgebirge Korsikas.– *Z. öst. EntVer.* **18**: 74-77.

Schille, F., 1912. *Oxyptilus* Z. *leonuri* Stange.– *Ent. Z. Frankf.a.M.* **26**: 103.

Schleich, C.L., 1864. Ueber die früheren Entwicklungsstände des *Pterophorus didactylus* Lin., Ev., (*trichodactylus* Hb.).– *Stettin. ent. Ztg.* **25**: 96-98.

Schmid, A., 1863. Beiträge zur Naturgeschichte der Schmetterlinge.– *Berl. ent. Z.* **7**: 65-66.

Schmid, A., 1886. *In*: Frey, H., 1886: Einige Micros aus Regensburg.– *Stettin. ent. Ztg.* **47**: 16-18.

Schwarz, R., 1953. *Motyli*, **III**. pp. i-vii, 1-159, 48 plates. Praha.

Scopoli, J.A., 1763. *Entomologia Carniolica exhibens insecta carnioliae indigena et distributa in ordines, genera, species varietates methodo Linnaeana.* 1-420, figs 1-815. J.T. Trattner, Vindobonae.

Sheldon, W.G., 1932. *Pselnophorus brachydactyla*, Kollar, as a British species, with notes on its earlier stages.– *Entomologist* **65**: 149-153.

Snellen, P.C.T., 1884. Microlepidoptera van Noord-Azie.– *Tijdschr. Ent.* **27**: 182-196.

South, R., 1881. Contributions to the history of the British Pterophorini.– *Entomologist* **14**: 49-53.

South, R., 1882. Contributions to the history of the British Pterophorini.– *Entomologist* **15**: 31-36.

Stainton, H.T., 1851. *A supplementary Catalogue of British Tineina and Pterophorini*. London.

Stange, G., 1882. Lepidopterologische Beobachtungen.– *Stettin. ent. Ztg.* **43**: 512-517.

Staudinger, O., 1857. Reise nach Island.– *Stettin. ent. Ztg.* **18**: 280-281.

Staudinger, O., 1859. Diagnosen nebst kurzer Beschreibungen neuer andalusischer Lepidopteren.– *Stettin. ent. Ztg.* **20**: 257-259.

Staudinger, O., 1870. Beitrag zur Lepidopteren Fauna Griechenlands.– *Horae Soc. ent. ross.* **7**: 279-283, Plate 3.

Staudinger, O., 1880. Lepidopteren-Fauna Kleinasien's.– *Horae Soc. ent. ross.* **15**: 159-435.

Stein, F., 1837. Entomologische Beobachtungen.– *Isis von Oken*: 98-109.

Stephens, J.F., 1835. *Illustrations of British Entomology: or, a synopsis of indigenous insects,...* etc.– Haustellata 4: 1-433. Baldwin & Cradock, London.

Strand, E., 1913. Weitere Schmetterlinge aus Kamerun, Gesammelt von Hern Ingeneur E. Hintz.– *Arch. Naturgesch.* **A.12**: 130-131.

Ström, H., 1783. Norske Insecters besskrivelte med Anmerkninger.– *Nye Saml. K. dansk. Vid. Selsk. Skr.* **2**: 49-93, 2 plates.

Sulzer, J.H., 1776. *Abgekürzte Geschichte der Insekten nach dem Linneischen System.*– Winterthur. 274 + 71 pp. 32 pls.

Sutter, R., 1988. *Stenoptilia annadactyla* sp. n.– *Reichenbachia* **25**: 181-184.

Sutter, R., 1991. Beiträge zur Insektenfauna der DDR: Lepidoptera-Pterophoridae.– *Beitr. Ent.* **41**: 27-121.

Sutter, R. & Skyva, J., 1992. *Stenoptilia stigmatoides* sp. n. aus der Slowakei.– *Reichenbachia* **29**: 81-82.

Svensson, I., 1966. New and confusing species of microlepidoptera.– *Opusc. ent.* **31**: 183-185.

Treitschke, F., 1833. *Die Schmetterlinge von Europa.* **9**: 1-294. Leipzig.

Turati, E., 1924. Spedizione lepidotterologica in Cineraica 1921-1922.– *Atti Soc. ital. Sci. nat.* **63**: 149-151.

Turati, E., 1926. Novita di lepidotterologica in Cirenaica.– *Atti Soc. ital. Sci. nat.* **65**: 66-67.

Turati, E., 1927. Novita di lepidotterologica in Cirenaica, II.– *Memorie Soc. ent. ital.* **66**: 313-344.

Tutt, J.W., 1896. A small collection of Lepidoptera from Lapland.– *Entomologist's Rec. J. Var.* **8**: 293.

Tutt, J.W., 1905. Types of the genera of the Agdistid, Alucitid and Orneodid plume moths.– *Entomologist's Rec. J. Var.* **17**: 34-37.

Tutt, J.W., 1907 (dated 1906). *A Natural history of the British Lepidoptera.* **V**. pp. i-xiii, 1-558. London & Berlin.

Wakely, S., 1935. Notes on captures during 1934.– *Entomologist* **68**: 177-180.

Walker, F., 1864. Lepidoptera Heterocera.– *List Specimens lepid. Insects Colln Br. Mus.* **30**: 926-953. London.

Wallengren, H.D.J., 1862. Skandinaviens Fjädermott.– *K. svenska VetenskAkad. Handl.* (N.F.) **3 (7)**: 1-25.

Wallengren, H.D.J., 1881. Genera nova Tinearum.– *Ent. Tidskr.* **2**: 94-97.

Walsingham, M.A., 1880. Pterophoridae of California and Oregon. pp. i-xvi, 1-66, 3 plates. London.

Walsingham, M.A., 1898. New Corsican microlepidoptera. – *Entomologist's mon. Mag.* **34**: 131.

Walsingham, M.A., 1908 (dated 1907). Microlepidoptera of Tenerife.– *Proc. zool. Soc. London*

1907: 911-926, plate 51.

Wasserthal, L., 1970. Generalisierende und metrische Analyse des primären Borstenmusters der Pterophoriden-Raupen.– *Zeitschr. zur Morphologie der Tiere* **68**: 177-254.

Whalley, P.E.S., 1960. The British Pyralidae and Pterophoridae in the Bowes collection, including a new species of plume moth.– *Entomologist's Gaz.* **11**: 29.

Whalley, P.E.S., 1961. *Pterophorus* Schäffer, 1766 (Insecta, Lepidoptera): proposal to place on official list Z.N.(S.) 1463.– *Bull. zool. Nom..* **18**: 159-160.

Wocke, M., 1864. Ein Beitrag zur Lepidopterenfauna Norwegens.– *Stettin. ent. Ztg.* **25**: 201-220.

Wocke, M., 1871. *In*: Staudinger, O. & Wocke, M.: *Catalog der Lepidopteren des europaeischen Faunengebietes.* Pp. i-xxxviii, 1-426. Dresden.

Wocke, M., 1898. *In*: Wernicke, H: Zwei neue, von Dr.M.Wocke beschriebene Microlepidopteren aus dem Groß-Glockner-Gebiet.– *Dt. ent. Z. Iris* **10**: 374-376.

Wolcott, 1936. Insectae borinquenses. A revised annotated check-list of the insects of Puerto Rico.– *J. Agric. Univ. P. Rico* **20**: 1-600.

Wolff, N.L., 1953. *Platyptilia capnodactyla* Zell., et zoogeografisk interessant fjermøl.– *Ent. Meddr.* **26**: 469-474.

Wolff, N.L., 1971. Tillæg og rettelser til fortegnelsen over Grønlands sommerfugle.– *Ent. Meddr.* **39**: 71-79.

Yano, K., 1963. Taxonomic and biological studies of Pterophoridae of Japan.– *Pacif. Insects* **5**: 65-209.

Zagulajev, A.K., 1986. Family Pterophoridae. – *In*: Medvedev, G.S. (ed.): *Keys to the insects of the European part of the USSR*, 4. Lepidoptera, **3**: 26-215.

Zagulajev, A.K. & Filippova, V.V., 1976. The new and little known species of plume-moths of the fauna of the USSR.– *Proc. Zool. Inst. Akad. Nauk USSR* **64**: 36-43.

Zeller, P.C., 1841. Vorläufer einer vollständigen Naturgeschichte der Pterophoriden, einer Nachtfalterfamilie.– *Isis von Oken* **10**: 756-794, 827-891, plate 4.

Zeller, P.C., 1847. Bemerkungen über die auf einer Reise nach Italien und Sicilien beobachteten Schmetterlingsarten.– *Isis von Oken* **12**: 895-910.

Zeller, P.C., 1850. Verzeichniss der von Herrn Jos. Mann beobachteten Toscanischen Microlepidoptera.– *Stettin. ent. Ztg.* **11**: 195-212.

Zeller, P.C., 1852. Revision der Pterophoriden.– *Linn. ent.* **1**: 319-413.

Zeller, P.C., 1867a. Skandinaviens Fjaedermott beskrifna af H.D.J. Wallengren besprochen.– *Stettin. ent. Ztg.* **28**: 321-339.

Zeller, P.C., 1867b. Einige von Herrn Picard Cambridge besonders in Aegypten und Palaestina gesammelte Microlepidoptera.– *Stettin. ent. Ztg.* **28**: 411-415.

Zeller, P.C., 1868. Beiträge zur Naturgeschichte der Lepidopteren.– *Stettin. ent. Ztg.* **29**: 401-429.

Zeller, P.C., 1873a. Beiträge zur Kenntnis der nordamerikanischen Nachtfalter, besonders der Mikrolepidopteren, I-III.– *Verh. zool.-bot. Ges. Wien* **23**: 201-334, 2 plates.

Zeller, P.C., 1873b. Lepidopterologische Beobachtungen vom Jahre 1872.– *Stettin. ent. Ztg.* **34**: 121-140.

Zeller, P.C., 1877. Exotische Microlepidoptera.– *Horae Soc. ent. ross.* **13**: 460-486, plate 6.

Zetterstedt, J.W., 1838-1840. Insecta Lapponica. vi + 1140 pp. Lipsiae.

Zimmerman, E.C., 1958. Lepidoptera: Pyraloidea.– *Insects Hawaii* **8**: i-xi, 1-456, ill. Honolulu.

Zukowski, R., 1960. Beobachtungen über das Vorkommen und Ökologie von *Aciptilia nephelodactyla* Eversm. in dem Pieniny-Gebirge (Polen).– *Polskie Pismo ent.* **30**: 243-252.

Index to hostplants

The numbers refer to the species number of the plume moths.

Index to entomological names

The index gives page reference to the key to genera, the check list, and the main treatment. Numbers in semibold refer to the valid taxa.

Printed in the United States
By Bookmasters